CRUSH HYPOTHESIS TESTING

THE 5-STEP METHOD TO MAKE HYPOTHESIS TESTING EASIER AND UNDERSTANDABLE

ALLISON DILLARD
JENNIFER FLENNER

HAPPY HYPOTENUSE PUBLISHING

Published by:

Happy Hypotenuse Publishing, LLC

Mission Viejo, CA 92691

Visit the author's website: www.allisonlovesmath.com

To contact the authors, please email: allison@allisonlovesmath.com

Crush Hypothesis Testing: A 5-Step Method to Make Hypothesis Testing Easier and Understandable / 1st ed.

Print ISBN: 978-1-950720-11-8

ALSO BY

Allison Dillard

Crush Math Now

Raise Your Math Grade

Love Math Journal

Available for purchase on Amazon.

For an updated list of titles, visit

www.allisonlovesmath.com

Allison
To Joe and my kids, for everything. My love for you is both statistically significant and practically significant.

Jennifer

To Arjuna for building a strong base; covid for creating urgency and space; and Allison for relentless drive and grace.

FREE BONUS BOOK

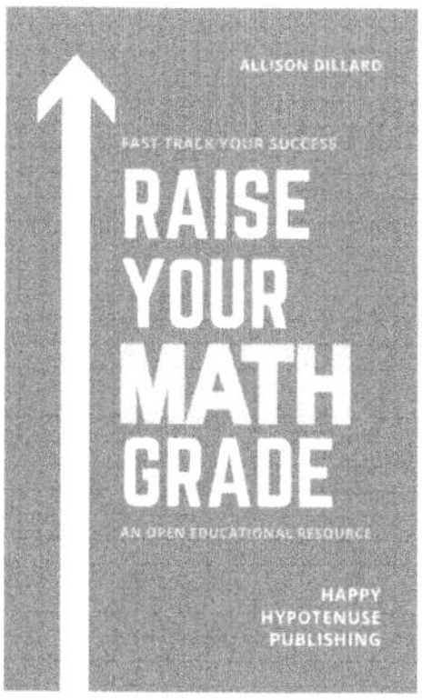

Raise Your Math Grade will teach you the mindset, study habits, and test taking skills necessary to succeed in difficult math classes, like Statistics. Download it for free at:

allisonlovesmath.com/free-book

CONTENTS

Introduction xiii

PART I
FOUNDATIONS

1. WHY HYPOTHESIS TESTING IS HARD 3
   14 Easily Confused Concepts 4
   Other Skills That Might Be Holding You Back 5
   A Bonus Life Skill 6

2. VOCABULARY YOU NEED TO KNOW 8
   Population and Samples 8
   Parameters and Statistics 10
   Proportion, Mean, and Standard Deviation 12

3. THE NORMAL DISTRIBUTION 14
   The Normal Curve 16
   Labeling Normal and Standard Normal Curves 17
   Normal Rule #1 Area = Probability 19
   Normal Rule #2 The Empirical Rule 20
   The Standard Normal Curve 21
   Converting Between the Normal and the Standard Normal 23
   Equations for Transformations 25

4. THE STUDENT T DISTRIBUTION 28
   Similarities Between Normal and Student t 29
   Degrees of Freedom 30
   Nine Differences Between the Normal and Student t Distributions 32
   How to Choose a Distribution 35
   Converting Between the Student t and Standardized Student t 36

5. CENTRAL LIMIT THEOREM 39
   What are Sampling Distributions? 40

6. CALCULATIONS YOU NEED TO KNOW 44
Finding Significance Levels 44
Calculating Critical Values 45
Calculating P-Values and Test Statistics 51
Using TI-84 Test Functions 53
Using Cumulative Distribution Functions (for P-Value) 54
Using a Formula (Test Statistic) 56

PART II

THE 5-STEP CRUSH HYPOTHESIS TESTING METHOD

7. THE 5-STEP METHOD 61

8. STEP 1 STATE THE HYPOTHESES 68
State the Parameter 68
State the Claim and Complement 69
State the Hypotheses 70

9. STEP 2 CHOOSE A DISTRIBUTION 74
Restate the Parameter 75
Check the Requirements 76
Choose a Distribution 76

10. STEP 3 SKETCH THE CURVES 78
Important Definitions 81
Label the Distribution, Units, and Mean 81
Label the Standard Deviation of the Curve 82
Shade the Significance Level Area 84
Label the Critical Value(s) and Rejection Region 84

11. STEP 4 DO THE TEST 86
Calculate the P-value and Test Statistic 87
Complete the Curve Sketch 89
Do the P-value Test 90
Do the Critical Value Test 92

12. STEP 5 STATE THE CONCLUSION 96
Follow the Flow Chart 96
Write the Basic Conclusion 97
Write the Meaningful Conclusion 98

PART III

HYPOTHESIS TESTING FOR A 1-SAMPLE POPULATION PROPORTION

13. THE 1-SAMPLE POPULATION PROPORTION QUESTION 103

Solution 104

14. STEP 1 STATE THE HYPOTHESES 111

State the Parameter 112

State the Claim and Complement 114

Write the Hypotheses 116

15. STEP 2 CHOOSE A DISTRIBUTION 118

Restate the Parameter 119

Check the Requirements 120

Choose a Distribution 124

16. STEP 3 SKETCH THE CURVES 126

Label the Distribution and Units 128

Label the Mean of the Curve 129

Label the Standard Deviation of the Curve 130

Shade the Significance Level Area 133

Label the Critical Values and Rejection Region. 134

17. STEP 4 DO THE TEST 140

Calculate the P-Value and Test Statistic 140

Complete the Curve Sketch 143

Do the P-Value Test 145

Do the Critical Value Test 148

Full Step 4 Solution 149

18. STEP 5 STATE THE CONCLUSION 157

Follow the Flow Chart 158

Write the Basic Conclusion 159

Write the Meaningful Conclusion 161

PART IV

HYPOTHESIS TESTING FOR A 1-SAMPLE POPULATION MEAN

19. THE 1-SAMPLE POPULATION MEAN QUESTION 165
Solution 166

20. STEP 1: STATE THE HYPOTHESES 172
State the Parameter 172
State the Claim and Complement 175
Write the Hypotheses 179

21. STEP 2 CHOOSE A DISTRIBUTION 182
Restate the Parameter 183
Check the Requirements 184
Choose a Distribution 188

22. STEP 3 SKETCH THE CURVES 190
Label the Distribution and Units 192
Label the Mean of the Curve 192
Label Standard Deviation of the Curve 194
Shade the Significance Level Area 197
Label the Critical Value and Rejection Region 198

23. STEP 4 DO THE TEST 203
Calculate the P-Value and Test Statistic 203
Complete the Curve Sketch 206
Do the P-Value Test 206
Do the Critical Value Test 209
Full Step 4 Solution 211

24. STEP 5 STATE THE CONCLUSION 222
Follow the Flow Chart 223
Write the Basic Conclusion 223
Write the Meaningful Conclusion 224

A+ Student Questions 227
CONCLUSION 229

About the Author 230
About the Author 231
Free Bonus Book 232

INTRODUCTION

As you make your way through the second half of your high school or introductory college Statistics class, you'll eventually learn about hypothesis testing. It's different from the math you likely saw in your other math classes and, if you're getting lost, feeling nervous, or have questions, you're not alone.

It's helpful to know up front that it's not just you. Hypothesis testing is the concept that stands between many students and an A (or passing grade) in Statistics. As Statistics teachers, we lose students every semester when we reach hypothesis testing. We'll get into exactly why that is soon, but for right now, we want to acknowledge that this is the topic that defeats many Introductory Statistics students. Don't worry, though. We're here to help!

We (Allison and Jennifer) are math professors who love teaching Introductory Statistics and have spent a pretty silly amount of time trying to simplify the complex and often confusing topic of hypothesis testing. Our goal with this book is to help you raise your grade (or keep it high) by better understanding one key concept: hypothesis testing.

Let's start with a quick overview to understand the big picture and why you're learning hypothesis testing in the first place.

What is Hypothesis Testing?

Suppose you're doing research—medical, political, psych, marketing, research on any topic, really. You're doing a survey or an experiment. You collect or are given a bunch of data. And then what? You get something that looks like this:

177.1600	178.5100	162.5200	165.9100
170.1700	172.8500	163.6400	157.2900
176.5100	155.4400	157.1000	168.1200
170.3500	162.5400	168.6400	187.0000
168.5000	171.5000	159.9900	157.2200

Tables like this aren't meaningful, so what do you do with the data? You calculate statistics, like the mean or standard deviation, to help make sense of it all.

A hypothesis test is simply a way to figure out if the statistics you've calculated are significant, meaningful, and relevant.

Another way to think of a hypothesis test is to look at the two words themselves. A *hypothesis* is an educated guess. A *test* is a formal procedure for determining the quality or performance of something. Hence, a hypothesis test is a formal procedure to determine if a hypothesis is of good quality, i.e., meaningful.

Hypothesis testing is arguably the most important and difficult concept in Introductory Statistics. That's why it's likely worth a large percentage of your grade and why we're writing an entire book about it to help you.

How to Crush Hypothesis Testing

To do well on hypothesis testing questions, follow the *Five-Step Crush Hypothesis Testing Method*:

1. State the hypotheses.
2. Choose a distribution.
3. Draw the graphs.
4. Do the test.
5. Write the conclusion.

The *Five-Step Crush Hypothesis Testing Method* is a method to help you master something difficult. As with all difficult things, it won't be easy the first few times you do it and you won't learn it just from reading about it. You need to practice to improve. Here's how that works for hypothesis testing:

1. Practice.
2. Make mistakes.
3. Learn from your mistakes.
4. Practice more.
5. Repeat until you can do the problems without error.

This is the process you must follow to learn all difficult things and the process you must follow to learn hypothesis testing.

PART I

FOUNDATIONS

Like all math, Statistics builds on itself. The concepts from the first part of Statistics are the foundations for the harder part—hypothesis testing. In this book, we don't cover *all* of Statistics. Instead, we've handpicked definitions and concepts that you need to crush hypothesis testing. Here are a few tips to help you make the most of Part I: Foundations.

Tip #1: Reflect on how you study. Use *1. Why Hypothesis Testing is Hard* to reflect on the challenge you're up against and what changes you can make to best prepare. Then, jump into studying.

Tip #2: Tackle it in 25-minute blocks. If you don't know the foundations, but your class is already on hypothesis testing, then you might feel overwhelmed by the amount of information you need to learn in a short period of time. If that's you, don't think about the big picture. Just set a timer for 25 minutes and go; you'll be amazed at how quickly you can catch up on the foundational concepts for hypothesis testing.

Tip #3: Remember that this book is *not* a comprehensive Statistics textbook. It's a supplement to your regular

Statistics text, not a replacement. The focus here is to develop material within the context of hypothesis testing and to clarify the details where you might get confused or lost. If you find that you need more information or examples, refer to your class textbook or class resources.

Ready? Let's get started!

1. WHY HYPOTHESIS TESTING IS HARD

Why on earth is Statistics so confusing? There are a bunch of reasons. Let's go through them, so you know what you're up against.

1. Hypothesis testing questions are word problems. You can't just plug them into an app that takes photos of homework questions and solves them for you. You have to understand and analyze the context and wording. We've created the *Five-Step Crush Hypothesis Testing Method* to make this easier for you.

2. Hypothesis testing requires new foundations. There may be some Statistics foundations (probability, normal distribution, sampling distributions) that you learned earlier in the semester that you've forgotten or didn't understand. We'll review these specific topics for you in the opening chapters.

3. There are many different hypothesis tests. Population proportion tests vs. population mean tests. 1-sample tests vs. 2-sample tests. Left tailed tests vs. right tailed tests vs. two-tailed tests. It's easy to get confused and do the wrong test or the wrong step within the right test. This is why we dive deep into only two tests: 1-sample hypothesis tests for population proportions and population means. Learn these two basic tests well, and the rest will become easier.

4. Everyone teaches it differently. Every YouTube video, teacher, and book explains how to do hypothesis tests slightly differently. Everyone does calculations with different tools and shows different steps in the process. Some teachers emphasize calculations, others, requirements, and others, interpretation. This is why each chapter lists common mistakes—things we and other teachers see over and over again when grading.

5. There are fewer people who can help you. For other math classes, you may have friends or parents who can help you. Statistics is often taken as a stand-alone class, so even if you have friends who aced Statistics, if it's been a while, they may not feel confident in their ability to help you. That doesn't mean you have to learn it alone. It means you might have to work harder to find someone to help you.

6. You learn hypothesis testing when you're burnt out. You'll likely get to hypothesis testing over halfway through the class when burnout is kicking in, when other classes are getting tough and time-consuming, and when it's hardest to force yourself to make and put in extra time. Make no mistake; burnout and lack of time are huge obstacles to success in hypothesis testing.

7. You might be approaching statistics with the wrong mindset. This is a problem for math in general, but especially for Statistics. People put math and Statistics up on a pedestal and talk relentlessly about how they are not good at math. If you're freaked out and approaching it with an "It's too hard! I'm just not a math person! I'll never figure it out!" mindset, then you're setting yourself up for failure before you've even begun. Toughen up. Math is just a skill. Practice, and you'll improve.

Those are the general big-picture reasons why hypothesis testing is hard.

14 Easily Confused Concepts

Here are fourteen specific points of confusion that lead to student mistakes in hypothesis testing. Use this as a checklist to make sure you know the foundations you need to know.

1. Samples vs. Populations
2. Statistics vs. Parameters
3. Proportions vs. Means
4. Normal Distribution vs. *Standard* Normal Distribution
5. Student t Distribution vs. *Standard* Student t Distribution
6. Normal Distribution vs. Student t Distribution
7. Sampling Distributions vs. Regular Old Distributions (and how their standard deviations differ)
8. Probabilities vs. Z-scores
9. Left-tailed vs. Right-tailed vs. Two-tailed tests
10. Null vs. Alternative Hypotheses
11. Significance Levels vs. Critical Values
12. Test Statistics vs. P-Values
13. P-Value Test vs. Critical Value Test
14. Wording for Conclusions

If you've mixed up any of these in a hypothesis testing question, you're not alone! Many students understand these concepts when reading the definitions but mix them up in the middle of a long, multi-step hypothesis testing question. This is why we cover these topics in the beginning *and* share common mistakes throughout the book.

Other Skills That Might Be Holding You Back

Students often struggle with Statistics for a bunch of reasons unrelated to math or how their class is being taught.

If you're struggling because you don't show your work, then

do your homework in our *Crush Hypothesis Testing Journal*. This will give you a dedicated place where the 5-Steps Method is already written out and can guide you through the work.

If you make algebraic mistakes, such as:

1. Simplifying calculations or equations
2. Drawing and labeling graphs
3. Converting between percentages, decimals, and fractions, or
4. Plugging numbers into equations

Then go back and review the Algebra skills that are leading to those mistakes.

Lastly, if you feel like you're struggling with hypothesis testing for any of the following reasons:

1. You are not a "math person."
2. Homework takes too long.
3. You don't have time for math.
4. You can't get motivated.
5. Math is too stressful or overwhelming.
6. You make careless errors.
7. You have math or test anxiety.
8. You're so far behind that you're failing.

Then download Allison's free book, *Raise Your Math Grade,* at https://www.allisonlovesmath.com/free-book.

A Bonus Life Skill

One of the side benefits of learning hypothesis testing is that you're learning how to do difficult things. Life will always have moments when it gets difficult, and you don't know if you have what it takes to succeed. Powering through hypothesis testing will teach you how to power through tough challenges, how to keep going in the face of uncertainty, and how to figure out how to do things that you don't know how to do. This bonus life skill of *powering through* is part of why different majors and jobs require Statistics. Employers use it to figure out, "Will this student give up when the work gets difficult and complex? Or will they power through until they've solved the difficult problem we hired them to solve?"

I guarantee that if you ever use hypothesis testing outside of school, you'll never see the exact questions from your Statistics class. You'll use the confidence you've gained from Statistics to look at equally or more complicated problems. You'll use your newfound ability to power through and solve difficult things.

Power through. With the *Five-Step Crush Hypothesis Testing Method* in your arsenal of tools to guide you, you have everything you need to make your way through the madness. Crush hypothesis testing today so that you can crush anything tomorrow.

2. VOCABULARY YOU NEED TO KNOW

Part of the confusion that comes with hypothesis testing is the language used. You learn so many new terms leading up to hypothesis testing that they are easy to forget or mix up. Students often chalk these errors up to "careless errors." Notice that I put "careless errors" in quotations because students insist that these mistakes are "careless errors," when they are often caused by insufficient knowledge of the prerequisite skills.

A few of the most basic mistakes students make are:

- Mixing up notation for statistics and parameters
- Mixing up notation for proportions and means
- Incorrectly calculating proportions, means, and standard deviations

Let's review these important concepts, so you don't get them mixed up.

Population and Samples

These two terms refer to a group of people, things, or data, but the difference between them is important.

Population: The *whole* group of people, things, or data values you're discussing or making a statement about. The population needs to be clearly defined.

Sample: A sample is a *subset* of the population. There are often too many people, things, or data values from a population to gather data, so we instead gather data from a sample.

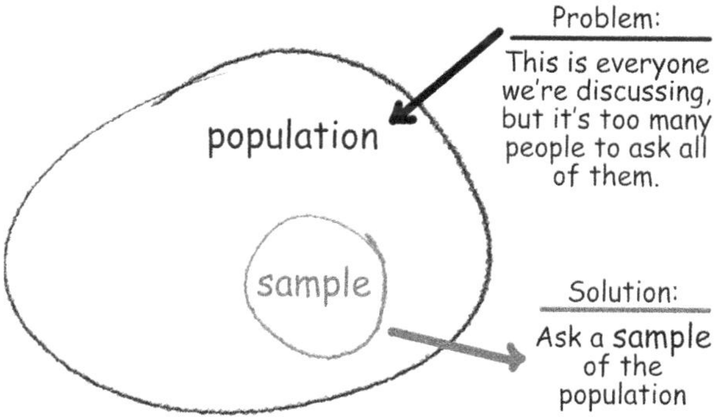

Keep track of what's a population and what's a sample in hypothesis testing questions. Here's how these two terms relate in a question we'll answer later in the book:

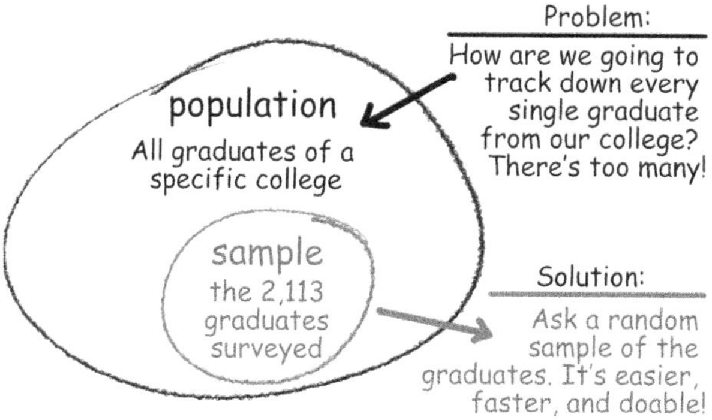

In this example, the population is all graduates of a certain college and the sample is the simple random sample of 2,113 graduates of the college that completed a survey.

Parameters and Statistics

This is where word problems get confusing. Words like mean, standard deviation, or proportion are used to describe data, but we need to understand the context of the problem to know if we are dealing with a population or a sample. Let me repeat that:

> The context of a word tells us if we are dealing with a population or a sample.

Why is this important? It's important because we do different things and use different notation depending on if we're dealing with a population or a sample.

A word (mean, standard deviation, proportion, etc.) that describes a population is called a parameter, and one that describes a sample is called a statistic. Remember *population parameters* and *sample statistics*. Population and parameter both start with the letter *p*; therefore, they go together. Sample and statistics both start with the letter *s*, so they also pair together. It's cheesy, but the alliteration helps students remember.

Parameter: A parameter is a word or calculation describing an entire *population*. Here are some important details about parameters:

- A parameter is a summary of population data.
- A parameter is rarely calculated directly because the population is too large to gather data from every single object or person in the population.
- Numerical values for parameters are generally either hypothesized based on prior knowledge or estimated from a statistic.
- Parameters can be represented in words: mean, proportion, standard deviation, size.
- Parameters can be represented with variables:

μ, p, σ, N

- μ = mean for a population
- p = proportion for a population
- σ = standard deviation for a population
- N = population size

Statistic: A statistic is a word or calculation describing a *sample*. Here are some important details about statistics:

- Numerical values for statistics are usually obtained by direct calculation.
- The numeric value of the statistic changes each time a new sample is drawn. A statistic varies from sample to sample.
- Statistics can be represented by words: mean, proportion, standard deviation, and size. (Notice: these are the same words used for parameters.)
- Statistics can be represented with variables:

$\bar{x}, \hat{p}, s, n$

- $\bar{x}$ = mean for a sample
- $\hat{p}$ = proportion for a sample
- s = standard deviation of a sample
- n = sample size

(Notice: the variables representing statistics are different from those representing parameters.)

Common Mistake: Thinking statistics and parameters are easy. After all, the definitions seem simple enough! The difficulty isn't the definitions themselves. The difficulty (and consequent mistakes) comes from students misunderstanding them in the context of a problem.

Use the correct variable notation for parameters and statistics on hypothesis testing questions; it matters. Here's a quick table to help you keep it all straight:

	population parameter	sample statistic
mean	μ	$\bar{x}$
proportion	p	$\hat{p}$
standard deviation	σ	s

One last warning: Whether you're discussing a parameter or a statistic, the *words* are the same, but the variables are different. The *context* of the word must be used to determine if it's a parameter or a statistic. Yes, one of the things that you need to succeed in hypothesis testing is strong reading skills. Dissect these problems as if you were in English class doing a reading comprehension assignment.

Proportion, Mean, and Standard Deviation

Hopefully, by the time you get to hypothesis testing, you know what the various parameters and statistics are, but if not, here's a quick recap:

Sample Size: This is the number of values, people, or data points in the sample. We use the variable n for sample size. (Note: We use the variable N for population size.)

Proportion: This is a fraction of the total that has a certain characteristic.

$$\text{proportion} = \frac{\text{Part}}{\text{Whole}}$$

We use the variable $\hat{p}$ for the sample proportion, and we use the variable p for a population proportion.

$$\hat{p} = \frac{\text{number w/characteristic}}{\text{sample size}} = \frac{\text{x in sample}}{n}$$

$$p = \frac{\text{number w/characteristic}}{\text{population size}} = \frac{\text{x in population}}{N}$$

Mean: This is the average of a data set. The mean is calculated by adding up all the values and dividing by the number of values.

$$\text{Mean} = \frac{x_1 + x_2 + \dots x_n}{n}$$

$$\bar{x} = \frac{x_1 + x_2 + \dots x_n}{n} \qquad \mu = \frac{x_1 + x_2 + \dots x_N}{N}$$

Standard Deviation: This is how spread out the data is. We use the variable *s* for a sample standard deviation and σ (pronounced "sigma") for a population standard deviation.

$$\text{standard deviation} = \sqrt{\frac{\Sigma\,(\text{observation} - \text{mean})^2}{\text{sample size}}}$$

$$s = \sqrt{\frac{\Sigma_{i=1}^{n}(x - x)^2}{n-1}} \qquad \sigma = \sqrt{\frac{\Sigma_{i=1}^{N}(x - u)^2}{N}}$$

3. THE NORMAL DISTRIBUTION

There are a bunch of probability distributions that you'll learn about in Statistics: Uniform, Bernoulli, Binomial, Poisson, Chi-Square, normal, Student t, F, and possibly more. The normal distribution is the single most important distribution for hypothesis testing in an Introductory Statistics class. It's a cute little graph that looks, roughly, like this:

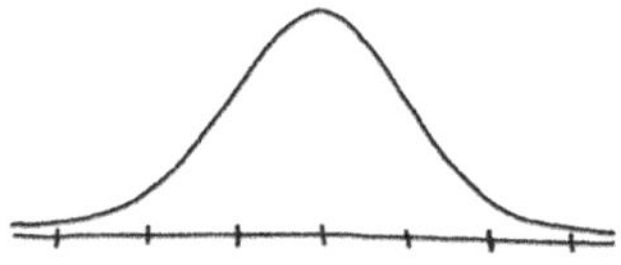

Here are a few things that define the normal distribution:

1. The graph is bell-shaped.
2. The graph is symmetric around the center (i.e., the mean).
3. The graph shows the number of standard deviations away from the mean.

In Statistics, the normal curve is important (i.e., it gets used a lot) because many data sets are normally distributed. For example, heights and weights are normally distributed. Test scores, like in your Statistics class or on the SAT, are normally distributed. Some financial data, like home prices or student debt loans, can be normally distributed. It comes up a lot, and that's why it's emphasized in Introductory Statistics classes.

For you, the student, the normal curve is also important because . . . it's a big part of your grade! So many homework and test questions use the normal distribution. Hypothesis testing, which likely makes up a large portion of your grade, requires the normal distribution. Every semester, I see overwhelmed students valiantly try to scrape by and pass Statistics without understanding the normal curve. They make so much more work for themselves and create so much more stress for themselves than is necessary; the shortcut (understanding the normal curve) is the easy and fast route to success in Statistics. To make sure you have a firm grasp of it, we recommend the following:

1. Read through our upcoming section on the normal distribution *and* the chapter on the normal distribution in your textbook.
2. Do *all* your homework on the normal distribution. So often, students skip this assignment without realizing just how much it'll mess them up later in the semester.
3. Ask your teacher (or someone else who can help you) if you have questions. If you haven't visited office hours yet but feel like your understanding of the normal distribution is fuzzy, this is a good time to get to know your teacher.

Before we dive in, I want to clear up a point of confusion that causes stress for many students—names for the normal distribution. The normal distribution goes by many names. If you're looking at different texts, blogs, videos, and lecture notes, all the following refer to the same thing:

- normal distribution,
- normal curve,
- normal probability distribution,
- Any version of the above with the word Normal (i.e., normal, but capitalized)
- bell curve,
- Gaussian distribution,
- Gaussian curve,
- Gaussian probability distribution.
- N(μ,σ)

Don't freak out. These are just different ways of referring to the normal distribution.

When we do 1-sample hypothesis tests, there are two different normal curves to keep track of—the normal curve and the standard normal curve.

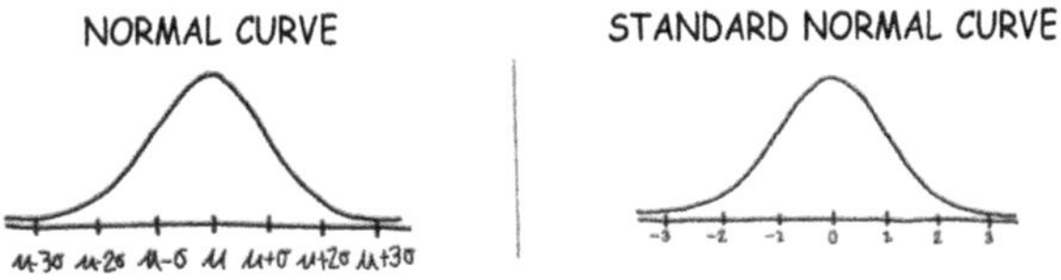

In this section, we'll review the normal and the standard normal curves and show you how to move back and forth between the two. In the section on 1-Sample Population Proportion hypothesis tests, we'll dive into how these curves relate to hypothesis testing.

The Normal Curve

Now, if you're one of the (many) students whose math anxiety went through the roof when your teacher first introduced the normal distribution, I want you to first think back to when you took Algebra. Even if it was a long time ago and you don't remember the details, you can hopefully remember that sometimes you drew things on graphs. Here are a couple graphs that you likely saw in Algebra, as well as the normal curve:

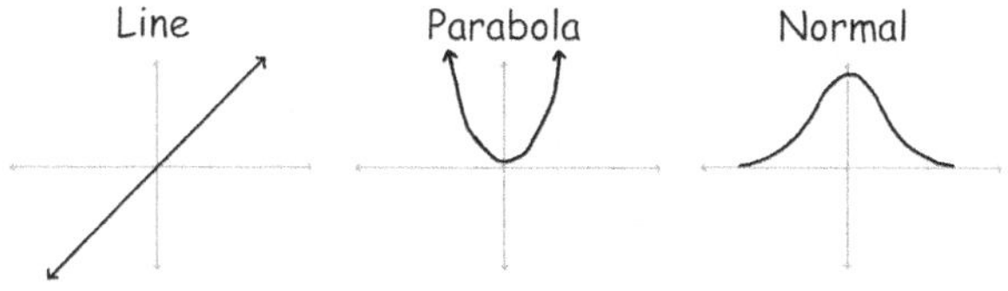

The normal curve is just like the graphs you saw in Algebra. Sure, it's a little weird looking. Yes, your teacher might introduce it along with a bunch of new intimidating terms, like probability density function, z-score, and area under the curve. Nevertheless, remember that in its most basic form, the normal curve is just a graph. It's no more special than a line or parabola.

Common Question: Is there an equation for the normal distribution? There sure is! Just as the general equation for a line is y=mx+b, the normal distribution also has an equation. It's pretty scary looking, though, and in many Introductory Statistics classes, you can get away with never looking at it, so we're leaving it out of this book for the sake of making everything less scary. If you want to know what the equation is, find it in your textbook and have fun playing around with it!

Labeling Normal and Standard Normal Curves

Just as the linear equation has certain important features (i.e., slope and y-intercept), the normal curve also has certain important features (i.e., the mean and standard deviation).

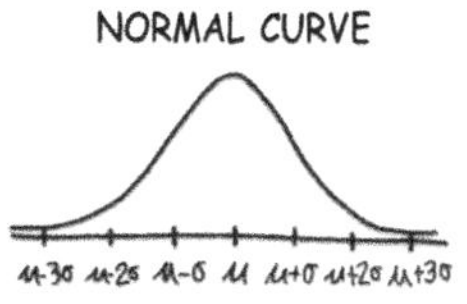

The graph above highlights the important features of the normal curve that are always the same:

- The curve is centered around the mean, μ.
- Each tick mark on the x-axis represents standard deviations, σ, above or below the mean (i.e., $\mu + \sigma$ is one standard deviation above the mean)

Let's look at how to draw a graph of a normal curve with an example. Suppose the mean of your data set is 0.5 and the standard deviation is 0.01088, draw the normal curve. (These are actual values from the 1-Sample Population Proportion problem we'll see later.)

For this particular normal curve, we can first label the mean, 0.5, at the center of the curve.

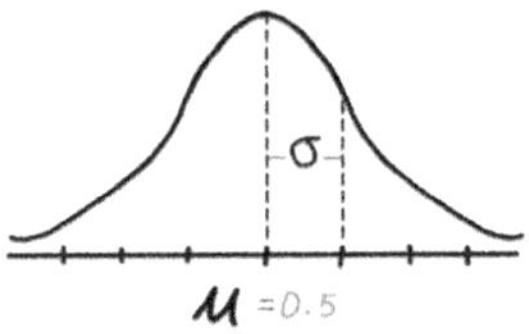

Next, we can find the values at the tick marks above and below the mean by adding or subtracting the standard deviation. These calculations look like this:

$$\mu - 3\sigma = 0.5 - 3 * 0.01088 = 0.4674$$
$$\mu - 2\sigma = 0.5 - 2 * 0.01088 = 0.4782$$
$$\mu - \sigma = 0.5 - 1 * 0.01088 = 0.4891$$
$$\mu = 0.5$$
$$\mu + \sigma = 0.5 + 1 * 0.01088 = 0.5109$$
$$\mu + 2\sigma = 0.5 + 2 * 0.01088 = 0.5218$$
$$\mu + 3\sigma = 0.5 + 3 * 0.01088 = 0.5326$$

Then, write these values in the appropriate place on the x-axis.

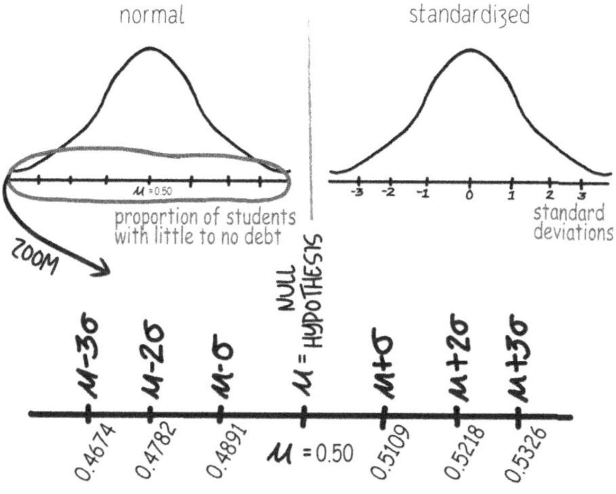

Normal Distribution Rules to Remember

Now that we know what a normal curve is and how to graph it, let's talk about the two most important things students overlook or don't understand about the normal curve. Both will help you better understand hypothesis testing and ace your Stats exams:

1. The area under the curve is a probability.
2. The normal curve follows the Empirical Rule.

Let's look at each in more detail.

Normal Rule #1 Area = Probability

The normal curve is a continuous probability density function, which means that the area under the curve adds up to one. Since probabilities range from zero to one, and any shaded area of the curve can only range from zero to one, there's a unique (and super convenient) relationship between area and probability.

For example, here's a graph we'll look at later.

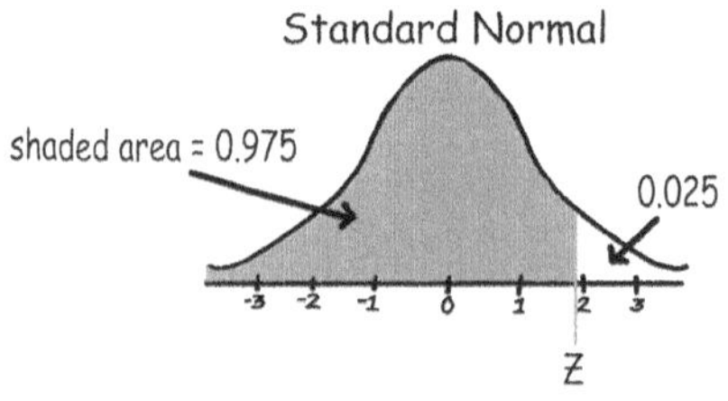

There are a few things we know from this graph:

- If the area to the right of z is 0.025, then the area to the left of z (shaded) is 1-0.025 = .975.
- If the area to the right of z is 0.025, then the probability of getting a value greater than z is 0.025.
- If the area to the left of z is .975, then the probability of getting a value less than z is .975.

This relationship between the area under the curve and probability is crucial to success in hypothesis testing.

Normal Rule #2 The Empirical Rule

In hypothesis testing, there's a lot of plugging things into your calculator. You're dealing with many inputs and outputs, and there are many places to misread, mistype, and generally make mistakes. The Empirical Rule will help you estimate what your answer should be so you can check your work. Consider it a commonsense check so you can tell when you accidentally plugged in the wrong number in your calculator. (This *will* happen at some point. There are just so many numbers to plug in.).

The Empirical Rule is best stated in terms of a graph (of course!):

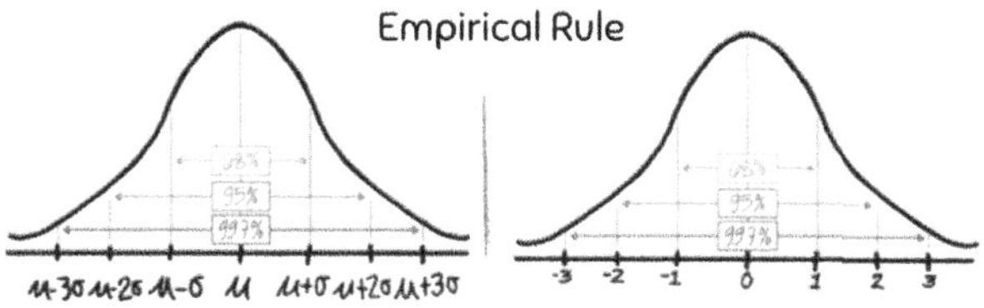

The Empirical Rule states that for the normal distribution:

- Approximately 68% of values fall within one standard deviation of the mean.
- Approximately 95% of the values fall within two standard deviations from the mean.
- Almost all values—about 99.7%—fall within three standard deviations from the mean.

Use the Empirical Rule to check your work and make sure your answers are in the right ballpark. It can help you to rule out bogus answers from incorrect calculator inputs.

Now that you know the most important characteristics of the normal distribution, let's take a look at the *standard normal distribution,* which is also important for hypothesis testing.

The Standard Normal Curve

The standard normal curve is the most basic normal curve. What do we mean by this? To understand, let's revisit those lines and parabolas from Algebra. Lines and parabolas, as well as other functions, have a most basic graph, and we call those graphs *parent functions*.

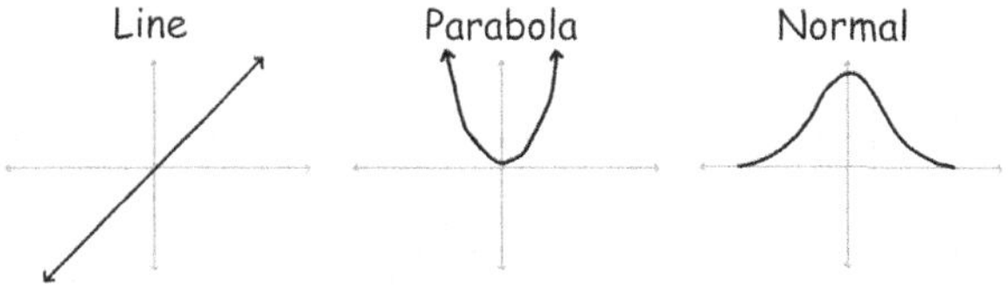

The graphs above are called parent functions. They can all be shifted up and stretched. Any version of these graphs that is shifted or stretched is made from this original most basic graph. This is how to think of the standard normal curve, as the parent function of all normal curves.

The standard normal curve has a few features that are important and always the same:

- The units for the x-axis are always standard deviations.
- The mean of the curve is always 0.
- The standard deviation is always 1.
- The z-score shows the number of standard deviations that a value is above or below the mean.

STANDARD NORMAL CURVE

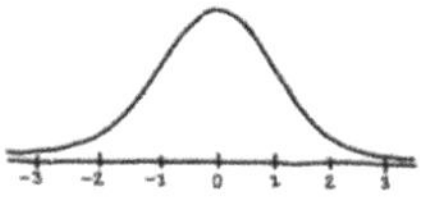

Common Question: How do we label the units of measurement on the standard normal curve? The values in the standard normal curve have been standardized, so we use z-scores, which represent the number of standard deviations away from the mean, as the unit of measurement on the x-axis. If your test question includes inches or dollars or degrees Celsius, the units for the standard normal curve are still z-scores.

Common Question: Why do we standardize the normal curve? The information you have to work with in any given hypothesis test question varies. Knowing how to move back and forth between the normal and standard normal curves is crucial for being able to work through the wide variety of wording and given information in hypothesis test questions.

The real reason is that standardization is a way to compare things that might have different units of measurement or a slightly different shape. Consider two different populations, each described by a normal curve. If these two normal curves have the same mean and standard deviation, then comparisons between the two populations are straightforward. However, how do we compare two data sets if they:

- have the same mean, but different standard deviations
- have the different means and the same standard deviation
- have different means and different standard deviations?

How on earth could we compare these different data sets? By converting (i.e., transforming) each to the standard normal.

Converting Between the Normal and the Standard Normal

To best understand transformations for normal curves, let's first look at something you might have seen in your Algebra class. Below are two graphs. The graph on the right is the parent function for linear equations, and the intercept is at the origin. The graph on the left is a different linear equation, and it does not go through the origin. It's a transformation of the parent function, the linear function. Different transformations of the linear function will have different y-intercepts and different slopes, but they all still transform from (and to) the parent function y = x.

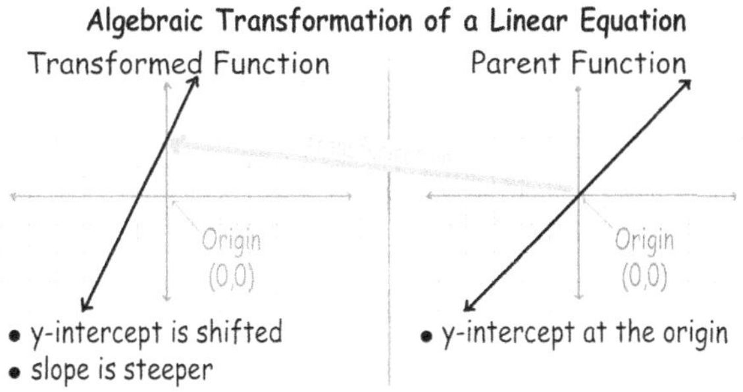

Similarly, below, the first graph is a transformation of the parent function, the quadratic function. Different transformations of the quadratic function will have different vertices and different axes of symmetry, but they all still transform from (and to) the parent function, the quadratic function.

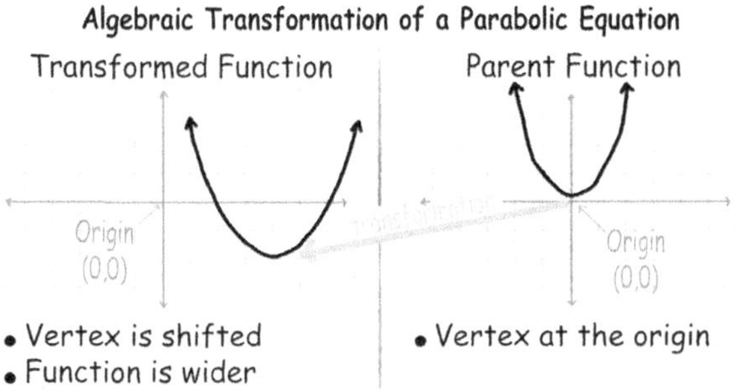

This is the same relationship that a normal curve has with the standard normal curve. The graph on the right is the standard normal curve, the parent function for all normal curves. The graph on the left is the transformed normal curve with a different mean and standard deviation.

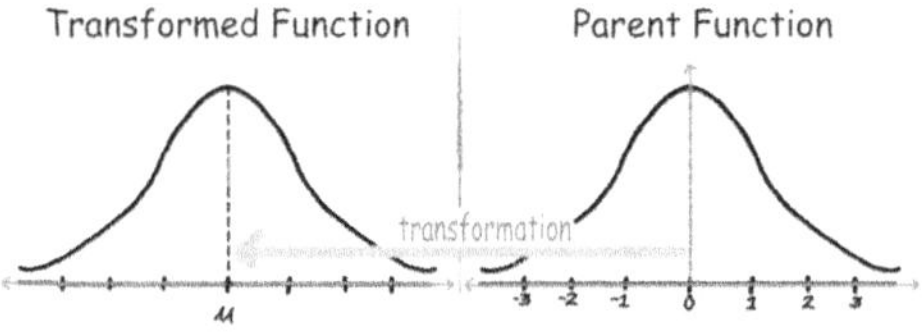

Like with the lines and parabolas in the previous examples, different transformations of the normal curve will have different means and different standard deviations, but they all still transform from (and to) the parent function, the standard normal curve.

Rather than drawing the graph in random places on a coordinate system, we zoom in on just the part of the graph needed to create the pair of graphs used throughout the book. For the details behind this, see the end of the chapter.

Equations for Transformations

Now that we know we can move back and forth between the normal and standard normal curve, let's talk about how to do it —equations.

Here is our first transformation equation, which allows us to transform the observation into a standard score. Given an observation on the normal curve (input), this equation finds the equivalent z-score on the standard normal curve (output):

EQUATION IN WORDS

$$\text{standard score} = \frac{\text{observation - mean}}{\text{standard deviation}}$$

Here's a sneak peek at how we'll use this for hypothesis tests about a population proportion:

The transformation above in words can be rewritten as an equation in variables for the 1-Sample Population Proportion normal curve:

- Standard score is Z
- Observation is statistic $\hat{p}$
- Mean is parameter p
- Standard deviation is $\sigma = \sqrt{\frac{pq}{n}}$

For the 1-Sample Proportion normal curve, this general equation becomes:

EQUATION IN VARIABLES

$$Z = \frac{\hat{p} - p}{\sqrt{\frac{pq}{n}}}$$

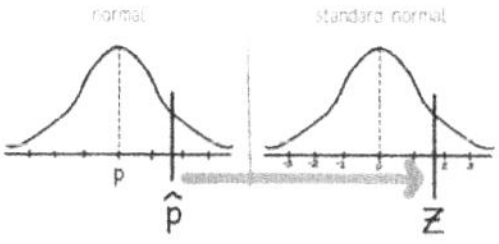

This equation allows us to take an observation on any normal curve and find the equivalent z-score on the standard normal curve.

We can also do a transformation in the opposite direction where we start with a z-score on the standard normal curve and find the equivalent observation on the normal curve. Given a z-score on the standard normal curve (input), this equation finds the equivalent observation on the normal curve (output):

$$\text{observation} = \text{standard score} * (\text{standard deviation}) + \text{mean}$$

It's important to notice that this is not a new formula! We use algebra to solve for the observation:

$$\text{standard score} = \frac{\text{observation - mean}}{\text{standard deviation}}$$

$$\text{standard score} * (\text{standard deviation}) = \text{observation - mean}$$

$$\text{standard score} * (\text{standard deviation}) + \text{mean} = \text{observation}$$

Again, we can fill this general formula in with variables for the 1-Sample Population Proportion normal curve to get:

$$\hat{p} = Z * \left(\sqrt{\frac{pq}{n}}\right) + p$$

Later, we'll also use this same formula to transform the *standard* critical value into an *observed* critical value:

Now, write the equation to explicitly transform a standard critical value, Z_α into an observed critical value X_α.

$$X_\alpha = Z_\alpha * \left(\sqrt{\frac{pq}{n}}\right) + p.$$

All three equations are doing the same thing; each transforms something from the x-axis of the standard normal curve onto the x-axis of the normal curve.

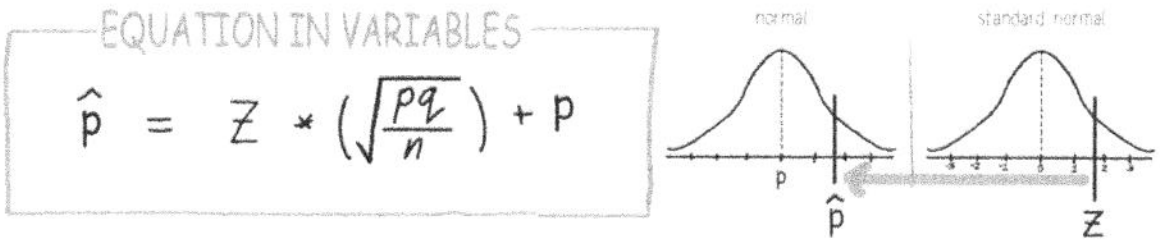

Common Mistake: Many students make mistakes when doing the order of operations. If you find yourself making mistakes with these calculations, look up PEMDAS or "order of operations" and do some practice problems. These Algebra foundations are crucial for success in Statistics.

A+ Student Question: Where do the transformations between normal curves and standard normal curves come from? This question requires a longer answer. Check out the answer, along with answers to other *A+ Student Questions,* at allisonloves-math.com/hypothesis-testing.

4. THE STUDENT T DISTRIBUTION

Before we jump in, let's clear up some confusion that causes stress for many students—names for the Student t distribution. Much like the normal distribution, the Student t distribution goes by many names. If you're looking at different texts, blogs, videos, and lecture notes, all the following refer to the same thing:

- Student t distribution
- Student's t distribution
- t distribution
- Student t curve
- Student t probability distribution
- Any variation of the above names with the word 'student' (i.e., Student, but not capitalized)
- Any variation of the above names with a hyphen (i.e., t-distribution, etc.)

Don't freak out because you are learning from people and resources that use different names. There are many slightly different ways to reference the Student t distribution. It's no big deal.

Names of the distribution are far from the most confusing parts of the Student t distribution but don't worry. We're going to clarify all the confusing stuff for you, and it's going to make hypothesis testing a heck of a lot easier!

In this chapter, we will:

1. Cover the similarities between the normal and Student t distributions.
2. Introduce some new concepts relating to the Student t distribution (degrees of freedom).
3. Create a list of differences between the two distributions.
4. Discuss how to use the Student t distribution in hypothesis testing and determine which distribution to use.

Similarities Between Normal and Student t

Let's dig into the similarities between the normal and Student t distributions.

At first glance, the Student t curve looks almost identical to the normal curve. The Student t distribution is very similar to the normal distribution. Very similar. Meaning, if you don't look closely, the two distributions will be almost identical.

Here is the Student t curve on the left and its standardized buddy on the right.

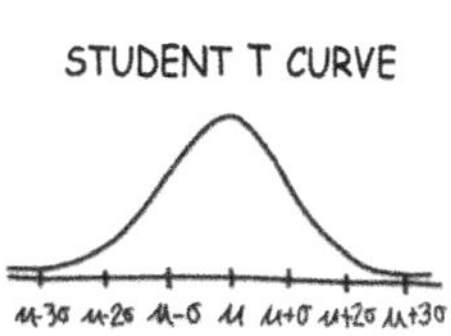

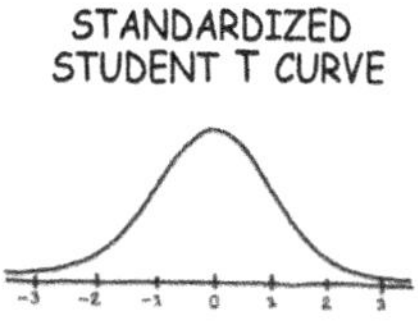

Wait, isn't that the normal curve we were looking at in the last chapter? Nope, this was the normal curve we saw in the last chapter:

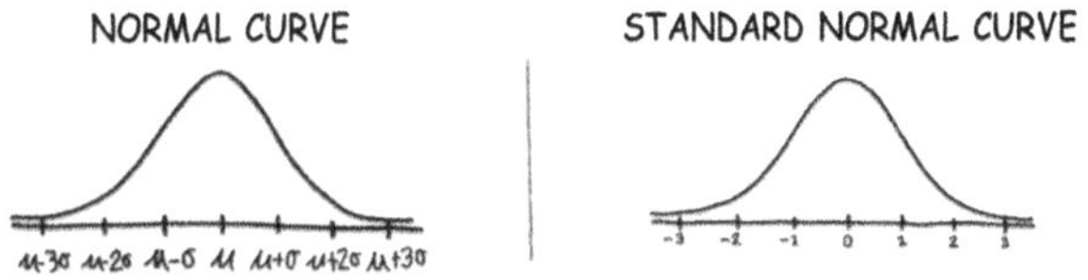

Yes, the two curves look similar! The similarity of how these two distributions look and function is a point of confusion for many students. If this is you, don't worry!

Here are a few of the things the *standardized* Student t and the *standardized* normal curves have in common:

Similarities Between Standardized Curves

	Standard Normal	Standardized Student t
shape	bell shaped, symmetric	
mean	centered around the mean, $\mu = 0$	
x-axis label	standard deviations away from the mean	
area under curve	total area under the curve sums to one	

Degrees of Freedom

The Student t curve was created because sometimes, we can't use the normal distribution. Sometimes, our sample sizes are too small, and we need a distribution that takes sample size into account. Hence, degrees of freedom!

Degrees of freedom are used to describe the number of values that are free to vary. For the Student t distribution to be used in a hypothesis test, we hypothesize or fix the population parameter at the center of the distribution. Once that parameter is hypothesized, there are *n-1* data points free to vary; the *n*th data point is constrained by the fixed parameter.

The shape of the Student t curve results from the number of data values that are free to vary; therefore, the shape of the Student t curve is defined by the degrees of freedom.

While we often label the Student t curve by the associated sample size, technically, the Student t curve is not defined by the sample size but by *degrees of freedom*, which will be calculated by subtracting one from the sample size.

Equation for Degrees of Freedom
df = n-1, where n is the sample size.

Each time the degrees of freedom change, the shape of the resultant Student t curve changes (ever so slightly). Each Student t curve is partnered with a distinct standardized Student t curve. There isn't a single standardized Student t curve, but rather a family of standardized Student t curves.

Let's look at these similarities and differences graphically:

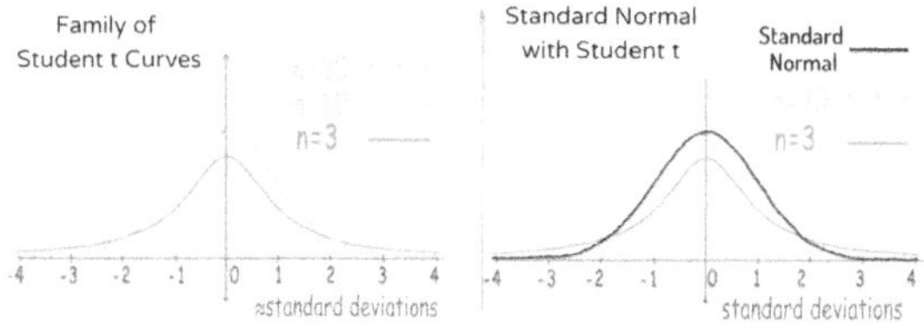

In the graph above, notice the width of the Student t curve depends on the sample size. A smaller sample size corresponds to a wider Student t curve. As the sample size grows, the Student t curve increasingly resembles the Standard normal. In fact, at sample size n = 30, the dotted line of the Student t curve appears to lay directly on top of the standard normal curve.

Common Question: Why are there multiple Student t curves but only one Standard normal curve? I'm glad you noticed! The Student t curve varies according to the sample size. For each sample size, we get a different Student t curve. Technically, this means the Student t curve isn't a single curve. It's a family of curves!

Important thing to note:

The Student t converges to normal for large samples.

As the degrees of freedom grow, the shape of the curve becomes increasingly similar to the standard normal.

Nine Differences Between the Normal and Student t Distributions

Here are the differences between the *standardized* forms of the two distributions that you'll want to be cognizant of:

1. *Defined by.* The normal curve is completely defined by its mean and standard deviation, while the standardized Student t curve is completely defined by the degrees of freedom. Because the curves are defined by different variables, they are generated by radically different mathematical equations.
2. *Notation.* Notation varies between the two distributions differ. The normal is denoted by N(μ,σ), while the Student t is denoted by T(df).
3. *Sample size.* There is only one standard normal curve, but there is a family of standardized Student t curves. Yes, even though when we sketch the graphs, they all look the same, there is a unique standardized Student t curve for each sample size or specific degree of freedom.
4. *Large sample.* As the degrees of freedom grow, the shape of the Student t curve becomes indistinguishable from the standard normal.
5. *Standard Deviation.* The standard deviation of the standard normal curve is always one; the standard deviation of the standardized Student t curve varies.

Each standardized Student t curve will have a unique standard deviation which will always be greater than one, yet close to one. For this reason, we will label the standard deviation for the entire family of standardized Student t curves as approximately one when sketches are drawn by hand.

6. *Tail size.* The standardized Student t curve has heavier tails than the standard normal. This means that more data is observed in the tail of the Student t distribution.
7. *Empirical Rule.* Technically, the empirical rule does not apply to the family of Student t curves. The empirical rule tells us that 0.3% of the data lies outside of 3 standard deviations on the normal curve; the tails of the standardized Student t curves contain even more data. (That being said, the differences are small, so Allison still likes to use the rule for approximating.)
8. *Standard scores.* The standard scores have a different name; on the x-axis of the standard normal curve, they are called z-scores, while on the x-axis of the standardized Student t curve, they are called t-scores.
9. *Choosing a distribution.* The two distributions are used in different situations. The choice of when to use each depends on the requirements that the problem statement satisfies—more on this in a bit.

Differences in notation between the two distributions are also important to keep in mind.

Differences Between Standardized Curves

	Standard Normal	Standardized Student t
standard score	Z-score	t-score
test statistic	Z	t
standardized critical value	$Z_{\alpha}, Z_{\alpha/2}$	$t_{\alpha}, t_{\alpha/2}$
distribution defined by	μ, σ	degrees of freedom df = n-1
standard deviation	$\sigma = 1$	$\sigma > 1$, but $\sigma \approx 1$
tail size	normal tail	fat tail

Having said all that, you probably have some questions:

Common Question: How can the area still be one if the width of the curve becomes wider? Great question! Height is a function of width. We know that (1) the area under the curve must equal one and (2) the Student t curves give us wider tails. For this to be true, then our height must be smaller. If you look at the graph, sure enough, this is true.

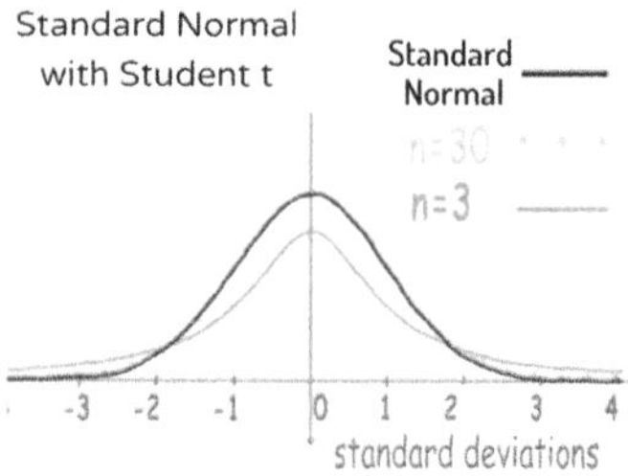

If you have a wider bell curve, but the area under the curve is the same, then the height must adjust accordingly.

Common Question: What is the equation for the Student t distribution? Teachers and textbooks rarely touch on this in Introductory Statistics. Google it, and you'll see why! It's a crazy equation. If you're a math or Statistics major or are simply curious, check it out! Otherwise, don't worry about it.

A+ Student Question: When talking about degrees of freedom, what do you mean by the number of values that are "free to vary?" Check out the answer, along with answers to other *A+ Student Questions*, at allisonlovesmath.com/hypothesis-testing.

How to Choose a Distribution

Choosing which distribution to use in your hypothesis testing problem is nontrivial and context-dependent. If you continue studying Statistics, you will learn more about the nuances involved in distribution selection. However, for an introductory course, you'll use the following rules:

Rule1 Choose A Distribution: population standard deviation

Problem about a proportion	Problem about a mean	
	*unlikely case σ known	σ NOT given s known
Normal	Normal	Student-t

*Unlikely case: The least likely real-world scenario is that you don't know the population mean, but somehow you know the population standard deviation. In this case, you'd use the normal. But that's wildly unlikely to happen, and these types of problems are frequently skipped over in Introductory Statistics classes.

Rule2 Choose A Distribution: sample size

n < 30	n $\geq$ 30
Student t	Normal

Caution: The Student t distribution should not be used with small samples from populations that are not roughly bell shaped.

It's important to always perform a thorough requirements check after selecting a distribution. This is so important in practice, yet so many students forget to do it.

Your teacher may give you a set of rules and requirements

that are more nuanced than what is written here, and you should use what your teacher provides, especially if it is different. However, for most Introductory Statistics courses, the rules provided here are all you need to know.

Converting Between the Student t and Standardized Student t

It's helpful to understand the transformations between a Student t curve and its standardized Student t curve. Just as you can use equations to move back and forth between the normal and standard normal curve, you use a similar equation to move back and forth between the Student t and standard Student t curves.

Here is our general transformation formula that we saw earlier. Given an observation (input), this equation gives you the standard score (output):

EQUATION IN WORDS

$$\text{standard score} = \frac{\text{observation} - \text{mean}}{\text{standard deviation}}$$

Next, we can plug in variables for the Student t curve.

For the Student t curve, we have:

- Standard score is t
- Observation is statistic $\bar{X}$
- Mean is parameter μ
- Standard Deviation is $\sigma = \frac{s}{\sqrt{n}}$

Therefore, for the Student t curve, this general equation becomes:

EQUATION IN VARIABLES

$$t = \frac{\bar{X} - \mu}{\frac{s}{\sqrt{n}}}$$

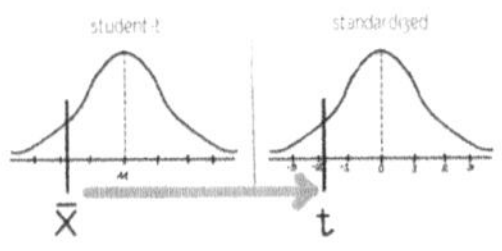

In a hypothesis test for a 1-sample population mean, this equation will allow us to use the observation on the Student t curve to find the test statistic on the standardized Student t curve.

Just as we saw with the normal distribution, we can also do transformations in the opposite direction. To do this, we can take the equation above and solve for the observation. Now, we have an equation that transforms the test statistic into an observation, something you may need to do in your homework for 1-sample hypothesis tests for population means.

EQUATION IN VARIABLES

$$\bar{X} = t * \left(\frac{s}{\sqrt{n}}\right) + \mu$$

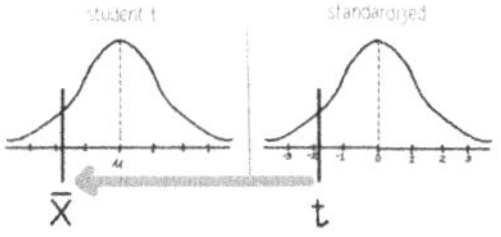

Another variation of this equation is also used when finding critical values for the Critical Value Test (more on that soon). You can rewrite this equation specifically with notation for the critical values:

$$X_{\alpha} = t_{\alpha} * \left(\frac{s}{\sqrt{n}}\right) + \mu$$

Students often think the following equations are all separate and unique, but when put together, we can see how they are just different ways of expressing the same transformation:

observation = standard score * (standard deviation) + mean

$$\bar{X} = t * \left(\frac{s}{\sqrt{n}}\right) + \mu$$

$$X_{\alpha} = t_{\alpha} * \left(\frac{s}{\sqrt{n}}\right) + \mu$$

All three equations transform something from the x-axis of the standardized Student t curve onto the x-axis of the Student t curve.

5. CENTRAL LIMIT THEOREM

We use the Central Limit Theorem (CLT) in two situations.

1. When we don't know some or all of the population parameters, the Central Limit Theorem can be used to learn parameters.
2. This second situation where we use the Central Limit Theorem is the important one, so we'll put it in a fancy box for you:

> When data has a non-normal distribution,
> the Central Limit Theorem allows us
> to convert that non-normal data into
> a sampling distribution of sample means (SDSM)...
> which is normal!!

We won't get too much into the details shown above. We could write an entire separate book on sampling distributions (SDSM) and the Central Limit Theorem (which sounds super fun . . . hmmm), but for right now, we'll streamline this to say: The CLT is super important for hypothesis testing! How?

Here's a flowchart to show you how to apply the CLT to hypothesis testing:

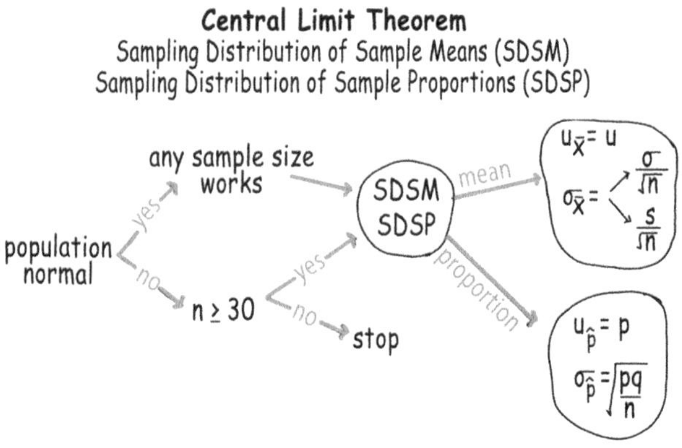

We need the CLT in hypothesis testing because it tells us the standard deviation of our curve. Depending on which distribution we use, we'll have the following:

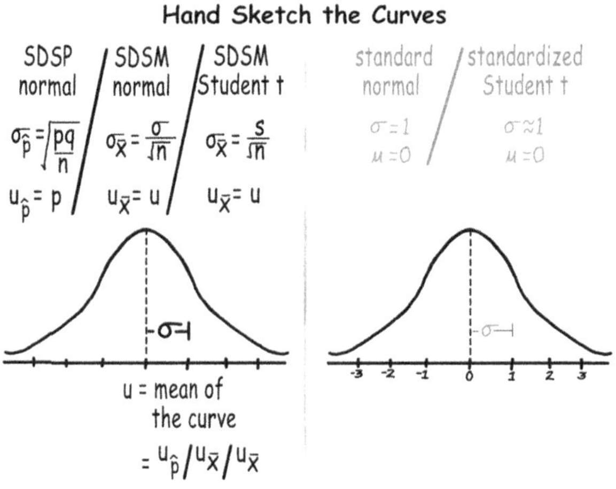

What are Sampling Distributions?

The terms SDSP and SDSM build on previous definitions you've learned in Statistics:

- Distribution - A complete list of values or data and

how often they occur.

- Probability Distribution - A distribution that describes the probability of each possible value occurring in a distribution.
- Sample - A subset of the population.
- Statistic - A word or calculation, like a mean or proportion, describing a sample.
- Sampling Distribution - A probability distribution of a statistic obtained from a larger number of samples drawn from a specific population.

Here's how the procedure of finding the Sampling Distribution of Sample Proportions (SDSP) works:

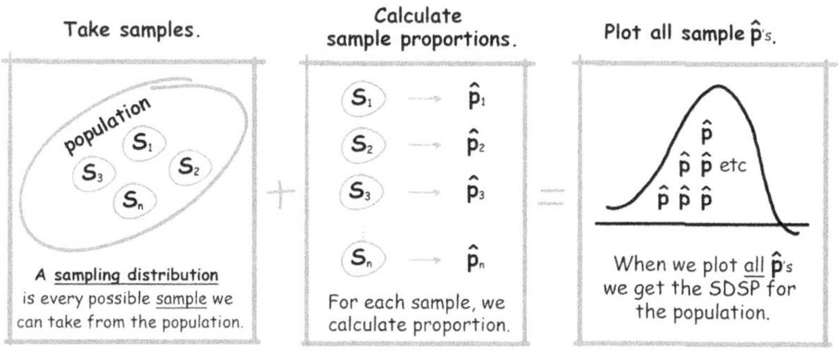

The procedure for finding the SDSM is exactly the same except that we're calculating the sample mean instead of the sample proportion, and the CLT applies.

To construct the SDSM, start by drawing many different samples from the population. Each of the drawn samples needs to be the same size; every sample must be size n. For every drawn sample, calculate the sample mean. Lastly, put all the sample means together and construct a distribution; this is the SDSM. The SDSM will be normally distributed even though the parent population is not! The key is that the samples must be large.

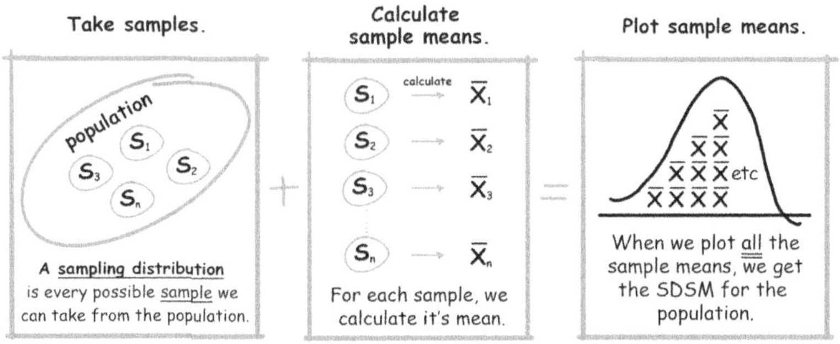

Now that we know the overview of CLT and a bit about sampling distributions, let's look at a formal definition of the Central Limit Theorem:

The Central Limit Theorem:
For all possible samples from a population,
of the same size n, where $n \geq 30$,
the sampling distribution of sample means (SDSM) can be
approximated by a normal distribution
with a mean, u, and standard deviation, $\frac{\sigma}{\sqrt{n}}$, such that,

$$N(u_{\bar{X}}, \sigma_{\bar{X}}^2) = N\left(u, \left(\frac{\sigma}{\sqrt{n}}\right)^2\right).$$

A more informal version of the Central Limit Theorem states:

1. If a population is *nearly* or *exactly* symmetrical, a *small* number of drawn samples will result in an SDSM with a normal shape.
2. If a population is not symmetrical, a *large* number of drawn samples is necessary for the SDSM to develop a normal shape.

Common Question: Is there just one sampling distribution per population or many? There are many possible sampling distributions for the same population. Each sample size n corresponds to a different SDSM, so one population has many possible SDSMs.

A+ Student Questions:

- How do sampling distribution of sample proportions (SDSP) and the CLT relate to sample proportions?
- How does CLT work for the Student t distribution?

Check out the answer, along with answers to other *A+ Student Questions,* at allisonlovesmath.com/hypothesis-testing.

6. CALCULATIONS YOU NEED TO KNOW

What's something that gets confusing for many students? You guessed it! Keeping the definitions straight for how to find:

- Significance levels and critical values
- P-values and test statistics

In this section, we'll go through the basic definitions of these concepts and clarify how to calculate them on a TI-84 calculator.

Finding Significance Levels

Before you learn about critical values, you must first understand significance levels. Here are a few important things to know about them:

- A significance level is the probability that an observation or test statistic provides significant evidence against the null hypothesis.
- It is denoted by the Greek letter α (pronounced "alpha").
- Because it is a probability, it represents an area on the normal or Student t curve. If you have a two-

tailed test, the area is split between the two tails. If you have a one-tailed test, the area falls within one tail.

How do we know if we have a one or two-tailed test? Our alternative hypothesis determines this:

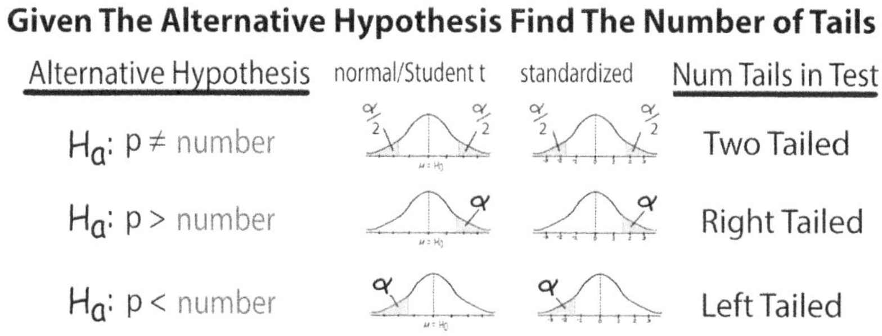

The alternative hypothesis determines where we shade **α** on our graph:

- In the third row, the alternative hypothesis is H_a: p < number, so we have a left-tailed test, and we shade **α** in the left tail of the curve.
- In the second row, the alternative hypothesis is H_a: p > number, so we have a right-tailed test, and we shade **α** in the right tail of the curve.
- In the first row, the alternative hypothesis is H_a: p≠ number, so we have a two-tailed test, and we shade **α**/2 in both tails of the curve.

Calculating Critical Values

The significance level is an area under the curve. Closely related to that is the **critical value**, a point on the x-axis that marks the end of the significance level (area under the curve) and the rejection region (on the x-axis).

The **rejection region**, also called the critical region, can be found by starting at the critical value and drawing an arrow pointing away from the mean. If an observation or test statistic is in the rejection region, then the null hypothesis is rejected. The image below shows the significance level areas, critical values, and rejection regions for a left tailed hypothesis test:

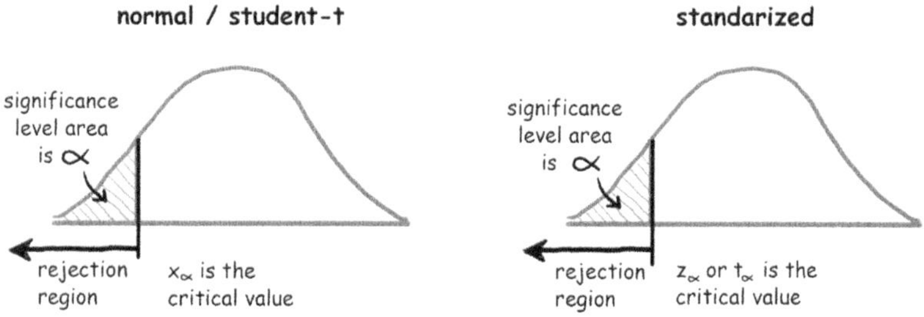

Common Question: What is the difference between *observed* and *standardized* critical values? Adding the words observed or standardized tells you which graph they belong on. In the figure above, the graph on the left is the observed space, and the graph on the right is the standardized space.

Common Mistake: Confusing significance levels and critical values. A significance level is an area, whereas critical values are numbers on a number line.

• α = the significance level. Significance level is an area.

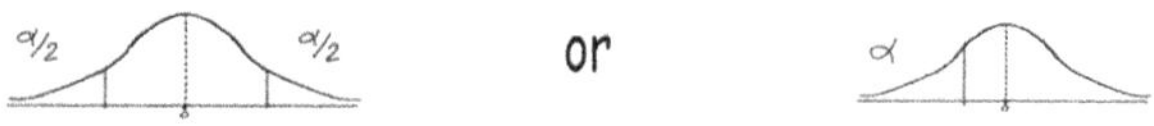

• z_α = the critical value. The critical value is a number on the x-axis of the distribution; on the standardized curve it is a standard score with an index related to the significance level.

Common Question: Why is my book's notation for critical values different from what I see here or online? Different textbooks, teachers, and resources use different notations for critical values. Some notation examples of critical value pairs for a two-tailed test include:

observed critical value notation:

$$(X_{\frac{\alpha}{2}}, X_{\frac{\alpha}{2}}), (X_{L\frac{\alpha}{2}}, X_{R\frac{\alpha}{2}}), (X_L, X_R), (X_1, X_2), \ldots$$

standardized critical value notation:

$$(-Z_{\frac{\alpha}{2}}, Z_{\frac{\alpha}{2}}), (Z_{L\frac{\alpha}{2}}, Z_{R\frac{\alpha}{2}}), (Z_L, Z_R), (Z_1, Z_2), \ldots$$

Our advice? Use whatever notation your teacher uses, and don't stress if other resources (like this book) use different notation.

To compute critical values for the normal distribution using a TI-84 calculator, use the invNorm (Inverse Normal Probability Distribution) function, which has three inputs:

- Area: This is the area to the left of the critical value—also called the significance level area.
- Mean: This is the mean of the curve. Look at your graph to find it.
- Standard Deviation: This is the standard deviation of the curve (not the data set).

Once you calculate the critical values, label them on the graphs along with the corresponding rejection regions.

Let's look at some examples. To start, find the critical value of a left-tailed test. Here's a graph you'll see later in the book:

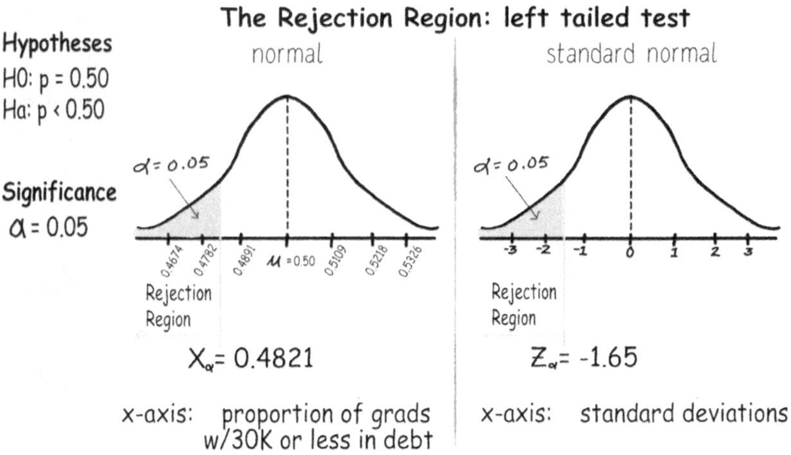

Given a significance level of .05, how do you find the critical value on the normal curve (the graph on the left)? Use the following inputs for the invNorm() function:

- area = .05,
- mean = 0.05,
- standard deviation = 0.019

Plugging in, you get:

$$x_\alpha = \text{invNorm}(.05, .05, 0.0109) = 0.4821$$

To find the critical value on the standard normal curve (the graph on the right), use the following inputs:

- area = .05,
- mean = 0
- standard deviation = 1

And you get:

$$z_\alpha = \text{invNorm}\ (.05,\ .05,\ 0.019) = -1.65$$

Common Question: Sometimes, the answer I get for my critical value is negative when it should be positive. Why is that? If this happened to you, you likely had a right-tailed test and didn't plug in the area to the *left* of the critical value. Let's look at an example:

Consider the following right-tailed test:

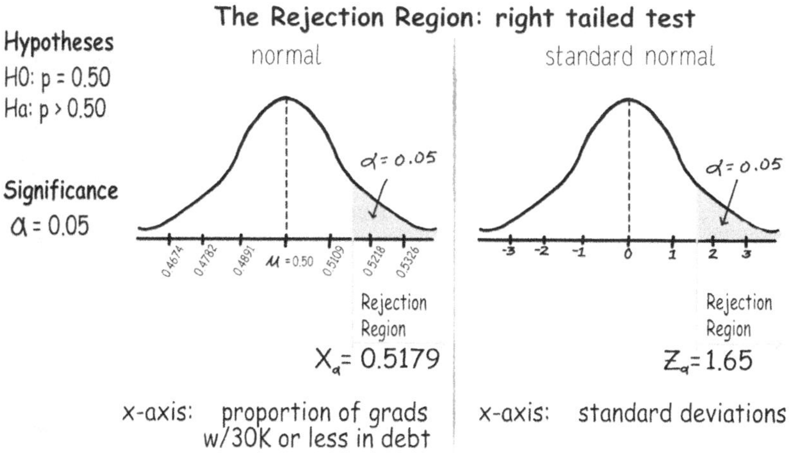

In the graph above, the area to the right of the critical value is 0.05, so the area to the left of the critical value is 1-.05 = .95. So, the correct way to calculate the critical value on the right tail is:

$$x_\alpha = \text{invNorm}\ (.95,\ .05,\ 0.0109) = 0.5179$$

On the standard normal curve, students often mistakenly plug in the area to the right:

$$z_\alpha = \text{invNorm}\ (.05,\ .05,\ 0.019) = -1.65$$

It's confusing because it's almost the correct answer, but it's negative. That, again, is caused by plugging in the wrong area.

Here is the correct calculation for the critical value for the right-tailed test:

$$z_{\alpha} = \text{invNorm}\ (.95, .05, 0.019) = 1.65$$

Common Question: Why do I always get the wrong answer for critical values when doing two-tailed tests? There are several mistakes you could be making in a two-tailed hypothesis test. First, you have to remember that the area in the tails is **α**/2, half of the significance level area (and not the whole significance level, **α**). Second, you need to remember that for the critical values on the right side, you still always plug in the area to the left of the critical value.

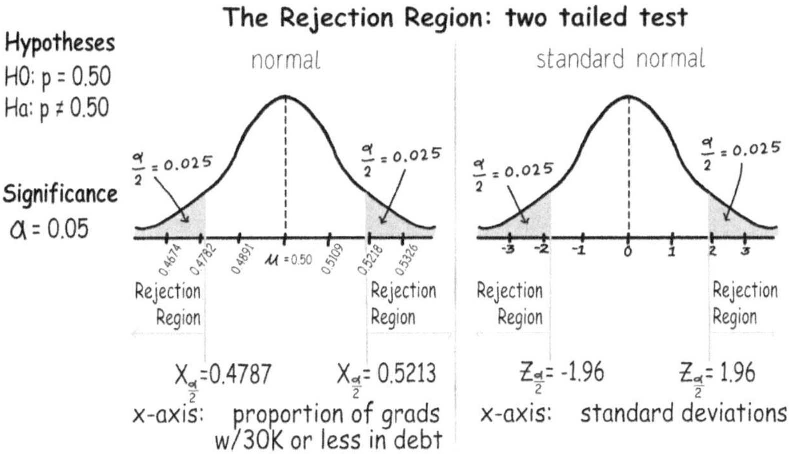

The correct calculations for the above critical values are:

$$x_{\alpha} = \text{invNorm}(.025, .05, 0.0109) = 0.4787$$
$$x_{\alpha} = \text{invNorm}(.975, .05, 0.0109) = 0.5213$$
$$z_{\alpha} = \text{invNorm}(.025, 0, 1) = -1.96$$
$$z_{\alpha} = \text{invNorm}(.975, 0, 1) = 1.96$$

Common Question: How do I calculate critical values for

the Student t distribution? When calculating critical values for the Student t distribution, we use the invT() function instead of the invNorm function. Since the Student t function is defined by degrees of freedom, the inputs are:

- Area to the left of the critical value
- Number of degrees of freedom.

Common Question: What do I do if I can't find the invT() function on my TI-83 calculator? This answer may frustrate you, but there is no invT() function on the TI-83 or TI-83+ calculator. For your homework, you can use one of the many free online tools that have the invT function; you might need to program the invT() function into your TI-83 calculator or purchase a TI-84 calculator. Talk to your teacher first, though, because if other students in your class also have a TI-83, they will be struggling with the same problem.

A+ Student Question: What's the difference between significance levels and confidence levels? Check out the answer, along with answers to other *A+ Student Questions*, at allisonlovesmath.com/hypothesis-testing.

Calculating P-Values and Test Statistics

In the 5-Step Method in the next section, you'll learn that there are two types of hypothesis tests: The P-Value Test and the Critical Value Test. We'll cover the details of the tests in the next section, but here, we want you to calculate two important things: p-values and test statistics. Here are the definitions:

- **P-value:** The probability of getting the value of the test statistic that is *at least as extreme* as the test statistic obtained from the sample, given that the null hypothesis is true. It is represented by an area under the curve to the left or right of the test statistic.

- **Observation:** A sample statistic observed in the real world with real-world units of measure. This observation will be placed on the normal or Student t distribution, and the x-axis will have units of measure.
- **Test Statistic:** A statistic we get when we convert the observation to a standard score. The test statistic goes on the x-axis of the standardized curve.

The image below shows where you will find the p-value, the observation, and the test statistic on the graph.

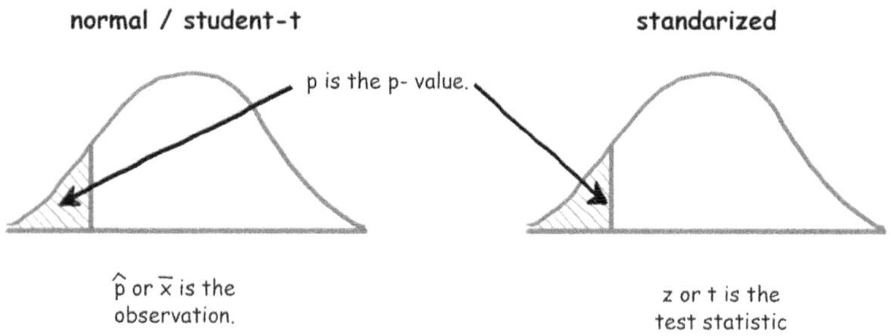

A few things to keep straight:

- The p-value is an area under the curve. It will be the same on both graphs.
- The observation is a statistic. It is found on the x-axis of the normal or Student t curve (the graph on the left).
- The test statistic is a standard score (i.e., a z-score or t-score). It is found on the x-axis of the standardized curve (the graph on the right).

There are three calculations you need to know for the p-value and the test statistic. We'll cover these in detail in the specific examples later, but for now, we want you to know what your options are:

Using TI-84 Test Functions

On your TI-84 calculator, press STAT, then scroll right to TESTS, and you'll see the following menu:

Getting There: STAT

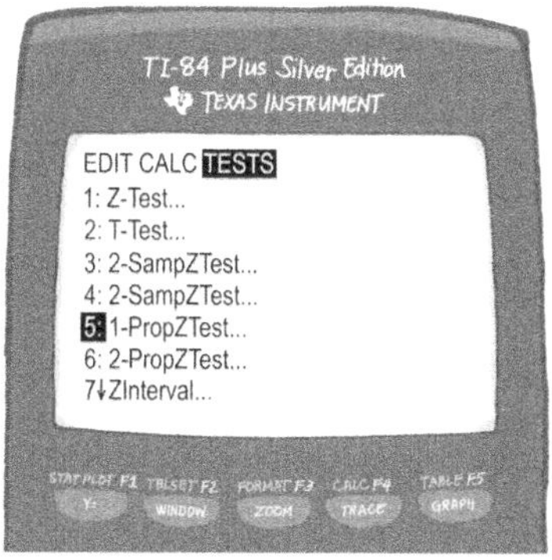

Which test you use depends on what hypothesis test you are performing:

- For hypothesis tests about proportions, choose 5:1-PropZTest.
- For hypothesis tests about a mean (with σ unknown, use 2:T-Test.
- In the unlikely event that you have a hypothesis test about a mean (with σ known), use 1:Z-Test.

The inputs will depend on your specific problem, but the cool thing about these test functions is they will give you both the p-value and the test statistic. (It gives you the observation also, but that's one of the inputs, so technically, you already have it.)

Common Question: Why does the notation we use in class

differ from the notation on my calculator? For some students, this was a point of confusion. See the notes below to see how the TI-84 calculator notation differs from the notation in your null and alternative hypotheses:

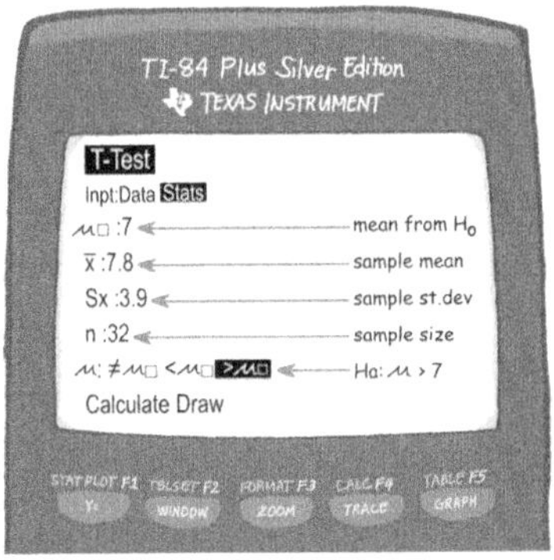

Common Question: Should I use the Z-Test for a hypothesis test about a proportion? So often, students gather their information about a population proportion and get stuck because the inputs don't match the information they've gathered! Why is that? It's because the Z-Test is actually used for hypothesis tests about a mean, when σ is known. This different (and rare) scenario also uses the normal distribution but has different inputs because it's a hypothesis test about a mean. Don't use the Z-Test for hypothesis tests about proportions. Use 5:1-PropZTest instead, and you'll find that the inputs will match the information you've gathered because it's the correct test.

Using Cumulative Distribution Functions (for P-Value)

Another way to find the p-value on your calculator is to press [2nd][VARS] to get the DISTR menu:

Which function you use again depends on what hypothesis test you are performing. To find the p-value for a 1-tailed hypothesis test:

- About a proportion, choose 2:normalcdf()
- About a mean (with σ unknown), use 6:tcdf()
- About a mean (with σ known), use 2:normalcdf()

To find the p-value for a 2-tailed hypothesis test, we need to remember that the *cdf* function (cumulative distribution function) finds the area under the curve in one tail, not two. Therefore, to find the p-value for a two-tailed test, we'll need to multiply by 2. To find the p-value for a 2-tailed hypothesis:

- About a proportion, choose 2:normalcdf() and multiply the answer by 2.
- About a mean (with σ unknown), use 6:tcdf() and multiply the answer by 2.
- About a mean (with σ known), use 2:normalcdf() and multiply the answer by 2.

Using a Formula (Test Statistic)

Remember, the test statistic is a standard score, meaning it's just a z-score or a t-score. Therefore, given an observation, we can use the formula for standardizing an observation to find our test statistic! Input the observation, mean, and standard deviation into the following formula to find the standard score:

equation in words

$$\frac{\text{observation - mean}}{\text{standard deviation}} = \text{standard score}$$

The output, the standard score, is the test statistic that we're looking for!

Of course, depending on which type of hypothesis test you have, you will write the observation and mean differently and calculate the standard deviation differently. Here's what the above equation looks like for each of the different hypothesis tests:

For a hypothesis test about a proportion, the equation to find a standard score is:

EQUATION IN VARIABLES

$$z = \frac{\hat{p} - p}{\sqrt{\frac{pq}{n}}}$$

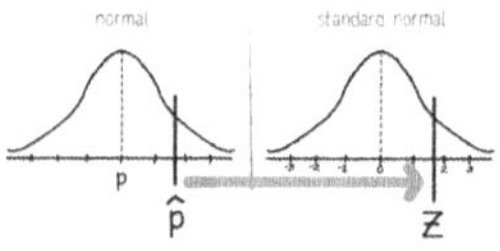

For a hypothesis test about a mean (with standard deviation unknown), the equation to find a standard score becomes:

EQUATION IN VARIABLES

$$t = \frac{\bar{x} - \mu}{\frac{s}{\sqrt{n}}}$$

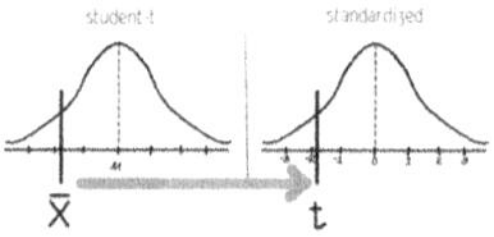

In the rare case that you have a hypothesis test about a mean (with standard deviation *known*), replace s with σ.

Now that you know how to find and where to graph critical values, p-values, and test statistics, let's jump into the *Five-Step Crush Hypothesis Testing Method*.

PART II

THE 5-STEP CRUSH HYPOTHESIS TESTING METHOD

What is a hypothesis test? The term hypothesis test breaks down into two words—hypothesis and test. A *hypothesis* is an educated guess about something based on limited information. In statistics, the hypothesis is a proposed claim about a population; it must be both quantifiable and testable. A *test*, in Statistics, is a procedure used to either reject or fail to reject the proposed claim.

A hypothesis test uses data from a sample to draw a conclusion about an entire population. It uses an observed sample statistic to make an inference about a population parameter.

Common Mistake: The first mistake students make in hypothesis testing is that they don't understand that a hypothesis test is *about a population.*

The mistake is so common that it's worth repeating:

> The hypothesis is a claim *about a population.* The hypothesis test checks a claim *about a population.*

Why is it the claim and test about a population, not a sample? Because you don't need to make claims or do complicated tests to check claims about samples. If you have a sample, you simply do the calculation. Done!

The problem lies with populations that are so large; you simply can't collect every data point. What's the average amount of student debt a student graduates with? There are just too many college graduates to collect data on every single one! Therefore, we must make a claim, use a sample, and conduct a hypothesis test to check the claim about that large population.

So, how do you check a claim?

It's gotta be tricky, right?

Not really, but for some reason, it often comes across trickier than it is.

7. THE 5-STEP METHOD

For that reason, we've broken hypothesis testing down into five steps that you can follow for every single hypothesis test:

1. State the Hypotheses.
2. Choose a Distribution.
3. Sketch the Curves.
4. Do the Test.
5. State the Conclusion.

It might be confusing the first few times because answering each step means that you need to understand each step, perhaps more thoroughly than you currently do. Once you've got it down, though, you'll know every step well; it'll start to feel repetitive, and you might actually consider hypothesis testing to be easy.

As we teach you how to organize and show your work with these five steps, you might show more work than your teacher requires. We do this for the following reasons:

- To get full credit on homework and test questions. You can never show too much work.
- To make sure that you understand what you're doing. Different videos, tutorials, books, and teachers show

different amounts of work. Students tend to get confused by what's left out—not by what's explained thoroughly.
- To clarify what your teacher is doing. In Algebra, everyone agrees on exactly what steps to show to solve an equation or graph a line. In Statistics, everyone teaches hypothesis testing differently, does calculations with different tools, and shows different amounts of work. By showing more than what your teacher might, we hope to fill in any gaps that are confusing you.
- To use the same steps in every single hypothesis test. We want the process to become repetitive and easy because that is when you know it well enough to crush your exam.

Here's a breakdown of the steps and charts you'll see in *The Five-Step Crush Hypothesis Testing Method.*

In Step 1, you'll determine which population parameter you are using to write out your hypotheses.

STEP 1: STATE THE HYPOTHESIS

Circle one: Proportion or Mean

Claim (in words): ______________________

____

Claim (in symbols): ______________________

Complement (in symbols): ______________________

Null Hypothesis: H_0 : ________ ______

Alternative Hypothesis: H_a ________ ________

In Step 2, you'll check that all the requirements are met in order to use the normal or Student t distribution. How much work you need to show here varies greatly from class to class. We've broken it down for you in case your teacher wants all the details.

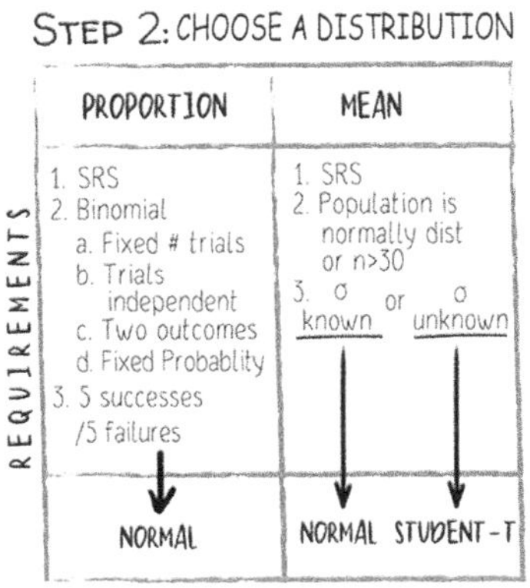

In Step 3, you'll draw the curves for your hypothesis test.

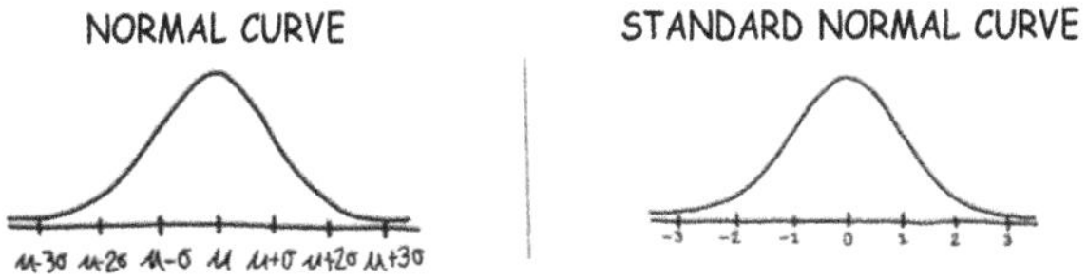

How do you go about filling in the correct information on the curves above? That is the typical confusing middle part of a hypothesis test where even the most ambitious students can get lost or make mistakes. To help, we created a checklist for you so you never miss a step and catch your mistakes.

Sketching Curves Checklist

	CHECKLIST	OBSERVATION SPACE	STANDARDIZED SPACE
1.	distribution ☐	normal / Student t	standard / standardized normal / Student t
2.	units (x-axis) ☐		NO CHOICE standard deviations
3.	mean of curve ☐	μ = Null Hypothesis	NO CHOICE $\mu = 0$
4.	standard deviation of curve ☐	circle one $\sqrt{\frac{pq}{n}}, \frac{\sigma}{\sqrt{n}}, \frac{s}{\sqrt{n}}$	circle one 1 or ≈ 1
5.	significance level α (area) ☐	same area, both curves α = ____	Left Tailed Right Tailed Two Tailed
6.	critical value (x-axis) ☐	$X_{\alpha}, X_{\alpha/2}$	$Z_{\alpha}, Z_{\alpha/2}, t_{\alpha}, t_{\alpha/2}$
	observation/ test statistic (x-axis) ☐	observation $\hat{p}$ or $\bar{x}$ =	test statistic Z or t =
	p-value (area) ☐	same area, both curves p-value = ____	Left Tailed Right Tailed Two Tailed

Here's a sneak peek at the completed curves for our 1-sample population proportion test in PART III:

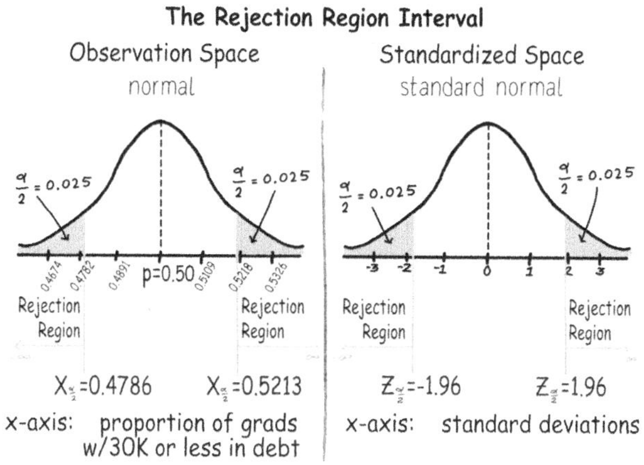

In Step 4: Do the Test. You'll calculate the observation, test statistic, and p-value, then use them to complete your curve and hypothesis test.

Sketching Curves Checklist

	CHECKLIST	OBSERVATION SPACE	STANDARDIZED SPACE
1.	observation/ test statistic ☐ (x-axis)	observation $\hat{p}$ or $\bar{x}$ =	test statistic z or t =
2.	p-value ☐ (area)	same area, both curves p-value = ____	Left Tailed Right Tailed Two Tailed

You'll add the observation, test statistic, and p-value to complete your curves. Here's a sneak peek at the completed curves for the 1-sample population mean test in PART IV:

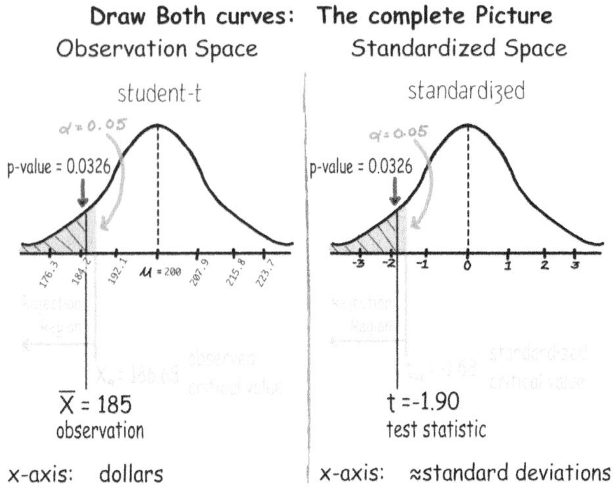

In Step 5, you'll use the P-Value Test or the Critical Value Test to complete the test and make your conclusion about the null hypothesis.

STEP 4: DO THE TEST

TEST	P-VALUE METHOD		CRITICAL VALUE METHOD	
CALCULATE	P-Value =		Test Statistic =	
TEST	Is P-Value ≤ α ?		Is Test Statistic in the Rejection Region?	
	No ↓	Yes ↓	No ↓	Yes ↓
DECISION	Fail to Reject H_0	Reject H_0	Fail to Reject H_0	Reject H_0

Lastly, in Step 5, you'll write up your final answer, being careful to choose the correct wording for your conclusion.

STEP 5: STATE YOUR CONCLUSION

Step 1: Claim	Step 4: Result	Conclusion
Claim is H_A	→ Reject H_0 →	There is sufficient evidence to support the claim that....
	→ Fail to Reject H_0 →	There is not sufficient evidence to support the claim that....
Claim is H_0	→ Reject H_0 →	There is sufficient evidence to warrant the rejection of the claim that....
	→ Fail to Reject H_0 →	There is not sufficient evidence to warrant the rejection of the claim that....

The goal of the *Five-Step Crush Hypothesis Testing Method* is not to make hypothesis testing easy. The goal is simply to clarify things that are difficult, fill in the gaps, and reduce your confusion, so you have everything you need to be successful at the difficult task of hypothesis testing.

Follow the steps. Follow them every time. Follow them regardless of the hypothesis test. Follow them until they get repetitive and obvious . . . because when the steps become repetitive and obvious, that's the sign that you understand hypothesis testing enough to crush your Statistics test.

In the upcoming chapters, we'll cover each of the steps in detail, including mistakes that students make and answers to the most common questions students ask. Read through the details. Take notes. Refer back for clarification when you get stuck.

8. STEP 1 STATE THE HYPOTHESES

Everything begins with the hypothesis. How you write your hypothesis will impact the rest of the problem. If the hypothesis is wrong, the entire problem is incorrect.

We've picked apart the first step so you can start strong and get your hypothesis correct. Here's what you need to fill in for Step 1: State the Hypotheses:

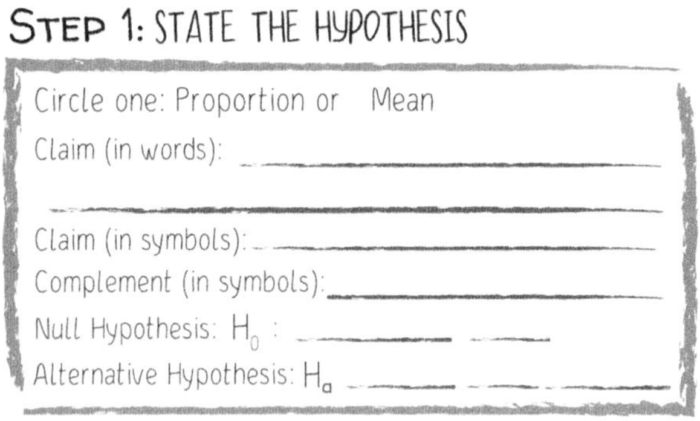

State the Parameter

What is the population parameter? To figure it out, read the question and look for keywords that indicate whether the question is asking about a mean or a proportion. Circle and label keywords and phrases.

Once you determine if the test will be about a proportion or a mean, identify the related population parameter in the chart.

	population parameter	sample statistic
mean	μ	$\bar{x}$
proportion	p	$\hat{p}$
standard deviation	σ	s

That parameter will be used to form the claim, the complement, and ultimately, the formal hypotheses.

Note: In this book, we'll cover only 1-sample population proportions and 1-sample population means; but in class, you might also do 2-samples or population standard deviations.

State the Claim and Complement

In real life, you will not necessarily be handed a nicely worded problem statement that presents a claim for you. A hypothesis may start with a vague statement or question, like "I think this new vaccine will work," or you'll be requested to run a test comparing "women's and men's salaries." In real life, determining your claim and finding or collecting sample data can be time-intensive steps. In your Statistics class, it's quicker.

In your Statistics class, your hypothesis is likely stated in a word problem. There isn't a lot of creativity or choice in the matter. Sometimes, the claim is written as a statement from a news source or an expert. It's just a matter of interpreting the word problem correctly—pulling out the words that tell you what the claim is. Highlight and copy the words verbatim.

After identifying the claim in a sentence or phrase, write it symbolically. If you have a 1-sample population parameter, there are six possible choices for the claim. The choices for 1-population proportions and 1-sample means are outlined in the chart below.

All Possible Claim Statements

Claim about a proportion	Claim about a mean
$p =$ number	$\mu =$ number
$p \leq$ number	$\mu \leq$ number
$p \geq$ number	$\mu \geq$ number
$p \neq$ number	$\mu \neq$ number
$p <$ number	$\mu <$ number
$p >$ number	$\mu >$ number

Once a claim has been determined, only one possible complement can be paired with it.

Claim and Complement Pairings

if this is the Claim	$\Rightarrow$	then this is the Complement
$p =$ number		$p \neq$ number
$p \leq$ number		$p >$ number
$p \geq$ number		$p <$ number
$p \neq$ number		$p =$ number
$p <$ number		$p \geq$ number
$p >$ number		$p \leq$ number

As you might remember from probability earlier in the semester, the complement of P(A) is 1-P(A).

The complement is the opposite. In the context of a hypothesis test, the complement is the statement that must be true when the claim is false.

State the Hypotheses

Next, we use the claim and complement to write the null and alternative hypotheses. Here are the different combinations of hypotheses for a test about population proportion:

All Possible Null & Alternative Hypotheses Combinations

Equality Statement	Null Hypothesis	Alternative Hypothesis
$p =$ number	$H_0: p =$ number	$H_a: p \neq$ number
$p \leq$ number	$H_0: p =$ number	$H_a: p >$ number
$p \geq$ number	$H_0: p =$ number	$H_a: p <$ number

Hypotheses for population means will be similar, but we'll use the variable for population means, μ, instead of population proportions, p.

Here are the definitions for each:

- **Null Hypothesis:** A statement that the parameter for a population equals something. You might hear your teacher refer to the null hypothesis as "H-sub-zero" (because H_0 has a zero subscript). The claim *or* complement can become the null hypothesis (whichever statement contains the equality).
- **Alternative Hypothesis:** A statement that the parameter is less than, greater than, or not equal to something. It will contain the symbols $<$, $>$, or $\neq$. The claim *or* complement can become the alternative hypothesis (whichever statement does not include equality). Depending on the book or resource you're using, the subscript for the alternative hypothesis might be A, *a*, or 1. In this book, we use the a subscript *a*.

How do we find the null hypothesis? One of the statements, either the claim or the complement, will contain an equality statement. We'll use the table below to create the null hypothesis, H_0:

The Equality Statement Becomes The Null

Equality Statement		Null Hypothesis
p = number	no change	H_0: p = number
p ≤ number	rewrite	H_0: p = number
p ≥ number	rewrite	H_0: p = number

Note: If the equality statement is not a strict equality (it's ≤ or ≥), it must be rewritten as a strict equality (=) to properly form the null hypothesis.

The "other" statement—the one that does not contain equality—will become the alternative hypothesis. The alternative hypothesis will never contain an equality statement; it will never be altered. It will always be written as it is initially formulated exactly.

You might hear your teacher talk about tests being right-tailed, left-tailed, or two-tailed. This is determined by the alternative hypothesis:

Given The Alternative Hypothesis Find The Number of Tails

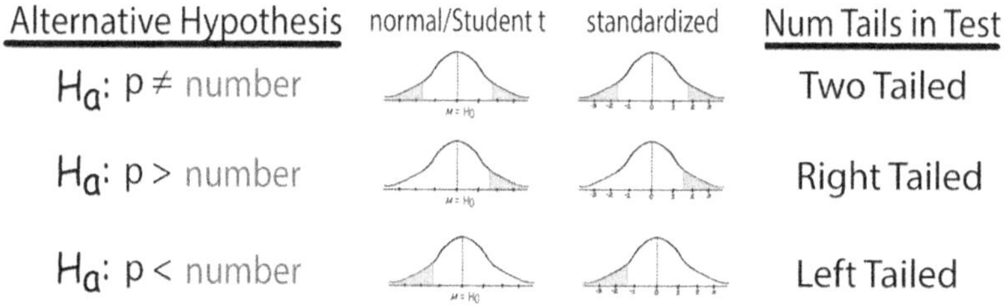

Common Question: Shouldn't my claim always be the null hypothesis? No! The claim is often not your null hypothesis. The null hypothesis is always an equality statement, even if your claim is not.

Common Mistake: Thinking the hypotheses are about sample statistics. Grr! Look . . . if we have a sample, we just calculate the sample proportion or sample mean. We don't need to do a hypothesis test if we have a sample. If we have a population, well, a population is big! We can't just track everyone down. It takes too much time, so use a sample and do a hypothesis test. A hypothesis test will always be about a population parameter.

If there's one piece of advice we can leave you with here, it's don't blindly plug in the claim or complement for either hypothesis. Start by remembering that the null hypothesis always has an equal sign in it.

9. STEP 2 CHOOSE A DISTRIBUTION

For a practicing statistician, the choice of distribution is nuanced and complicated. For an Introductory Statistics course, the act of choosing a distribution is simpler but still confusing. We've broken it down into what we think is the clearest and most straightforward way, but this is one of those situations where different texts, teachers, and online resources all use slightly different wording, charts, and tables.

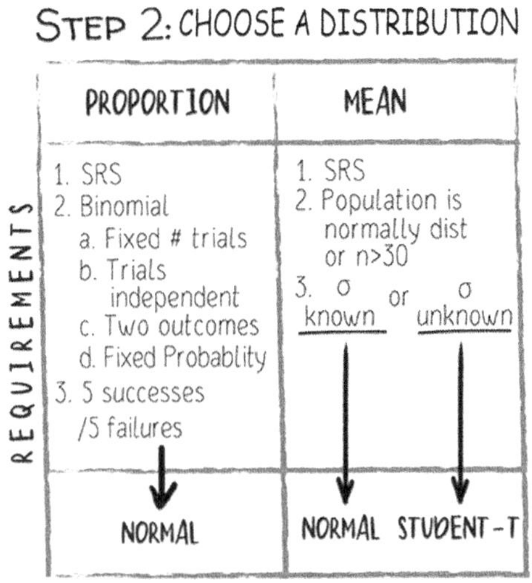

Common Question: That's not how I was taught to choose a distribution! Is there another way? Yes. If the way we break this down doesn't look similar to what you learned in class, follow whatever rule your teacher or text gives you for choosing a distribution. Given a problem about a population mean, you can choose a distribution: (1) based on the population standard deviation and (2) based on the sample size.

If your teacher chooses the distribution based on standard deviation, the rule you follow might look like this:

Rule1 Choose A Distribution: population standard deviation

Problem about a proportion	Problem about a mean	
	*unlikely case σ known	σ NOT given s known
Normal	Normal	Student-t

If your teacher chooses the distribution based on sample size, the rule you follow might look like this:

Rule2 Choose A Distribution: sample size

$n < 30$	$n \geq 30$
Student t	Normal

Caution: The Student t distribution should not be used with small samples from populations that are not roughly bell shaped.

While these rules certainly look confusing, they are very similar—just emphasizing different things.

Restate the Parameter

Below, we've condensed the requirements for choosing a distribution into a checklist for you. In the top row, circle your population parameter.

Check the Requirements

The second row tells you which requirements to check given your parameter.

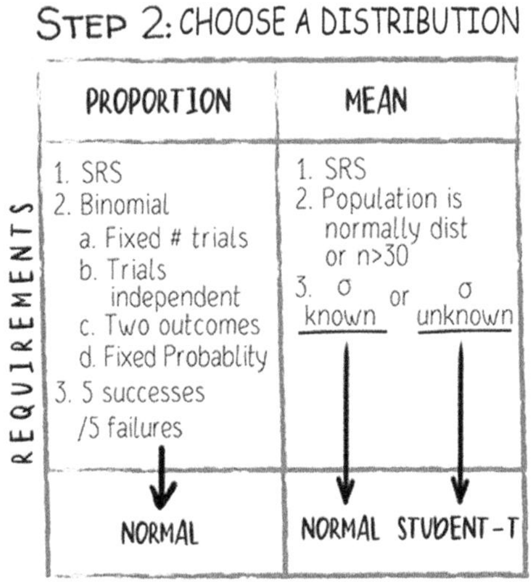

It's important to remember that it is not appropriate to perform a hypothesis test if the requirements are not met. Verify that the requirements are met before selecting your distribution.

Common Question: Can I just skip the requirement checks? Pay attention to what your teacher wants. Some blogs and YouTube videos skim or skip over this, but if your teacher wants you to check the requirements and write them out for each problem (like we do), then you better include them if you want full credit.

Choose a Distribution

If the requirements are met, circle the corresponding distribution, and you're done! Then move to Step 3: Sketch the Curves. Before we jump into that, though, there are a few mistakes and questions that come up when choosing a distribution.

Common Mistake: Using the normal distribution when you should use the Student t distribution. We use the normal distribution when we're dealing with population proportions. We use the Student t distribution (usually*) when we're dealing with population means. Yes, there's a (*) there for a reason. When we're doing a hypothesis test about a population mean, we don't usually know the population standard deviation. On the rare chance that the population standard deviation is known, but the population mean is not, you'll use the normal distribution.

10. STEP 3 SKETCH THE CURVES

A picture is worth a thousand words, and drawing a picture is the key to correctly answering hypothesis testing questions. The goal of Step 3 is to sketch two curves so that you will get a complete picture of the problem and catch mistakes in calculations. The sketch is not intended to be perfect, so don't let that slow you down.

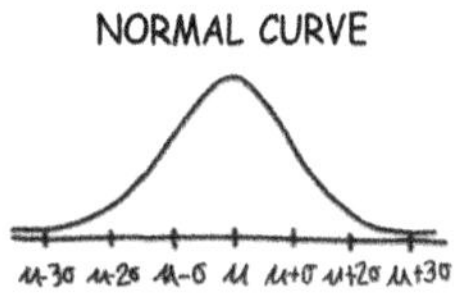

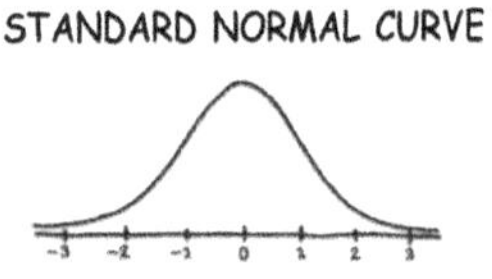

Sketching both curves is important, so don't skip this step. While it can confuse students, we've broken it down into a checklist for you! Use it to make sketching the graphs straightforward:

Sketching Curves Checklist

	CHECKLIST	OBSERVATION SPACE	STANDARDIZED SPACE
1.	distribution ☐	normal / Student t	standard normal / standardized Student t
2.	units (x-axis) ☐		NO CHOICE standard deviations
3.	mean of curve ☐	μ = Null Hypothesis	NO CHOICE $\mu = 0$
4.	standard deviation of curve ☐	circle one $\sqrt{\frac{pq}{n}}$, $\frac{\sigma}{\sqrt{n}}$, $\frac{s}{\sqrt{n}}$	circle one 1 or ≈ 1
5.	significance level α (area) ☐	same area, both curves α = ____	Left Tailed Right Tailed Two Tailed
6.	critical value (x-axis) ☐	x_{α}, $x_{\alpha/2}$	z_{α}, $z_{\alpha/2}$, t_{α}, $t_{\alpha/2}$
	observation/ test statistic (x-axis) ☐	observation $\hat{p}$ or $\bar{x}$ =	test statistic z or t =
	p-value (area) ☐	same area, both curves p-value = ____	Left Tailed Right Tailed Two Tailed

The checklist breaks down the six things you need to label on the graph:

1. Label the distribution.
2. Label the x-axis units.
3. Find and label the mean of the curve.
4. Calculate and label the standard deviation of the curve.
5. Determine the number of tails and shade the significance level area.
6. Find and label the critical value(s).

The worst mistake a student can make is thinking, *My teacher doesn't make me show these two graphs, so I'll skip this step.* But you should get in the habit of creating the graphs for the following reasons:

1. You'll have a deeper understanding of hypothesis testing, so you can maneuver through the entire hypothesis testing problem without getting overwhelmed or lost.
2. You can figure out how to show your work the way your teacher wants you to show it, which will result in higher grades.
3. You'll be able to spot and work through hard, tricky test questions. (I call these test questions *curveballs,* and I always include at least one curveball on tests to differentiate between students who deserve an A and those who deserve a B. If you want an A in Statistics, be ready to earn it!)
4. You'll be able to make sense of teachers, books, and resources that explain hypothesis testing solutions in different ways.
5. Most importantly, you'll be able to check your work. There is a mapping between the two curves. If you fill in everything correctly, both graphs lead to the same conclusions. If you make a mistake, you'll see it right away because the curves won't match. It's particularly important when using technology to have a method to check your work. One typo can throw everything off.

All that said, use your judgment. Learn what you want and need, which depends entirely on you, your class, and the level of understanding you are striving for. If your teacher does not require you to show these two curves, then pay attention to which curve she uses and make sure to fill in at least the one curve that she does use.

The truth is you only need to place all the information on one curve to do a hypothesis test. If you fill in everything correctly, both graphs lead to the same conclusions.

Important Definitions

There are some important definitions that you need to know for Step 3. We covered these earlier, but here is a recap:

- The **significance level**, α, is an area under the curve that represents the probability that an observation or test statistic provides significant evidence against the null hypothesis.
- The **critical value** is a point on the x-axis of a distribution that marks the end of the significance level area.
- The **observed critical value** is a critical value in the observation space (the graph on the left).
- The **standardized critical value** is a critical value in the standardized space (the graph on the right).
- The **rejection region** (also called the critical region) is an interval on the x-axis. It can be found by drawing an arrow along the x-axis, starting at the critical value and pointing away from the mean. If an observation or test statistic is in the rejection region, then reject the null hypothesis.

Let's go through the steps of the checklist in detail.

Label the Distribution, Units, and Mean

1. Label the distribution.

We first label the distribution. Is it normal or Student t? Look back at Step 2 to see which requirements were met and which distribution you'll use.

2. Label the x-axis units. Teachers always want you to label units on word problems, and hypothesis testing is no different. Look back at the context of the question to determine your units. What are the units of measurement of the mean or proportion? Do you have dollars, pounds, inches? The units of measurement related to the null hypothesis will be the units of measurement for the x-axis on the normal or Student t curve. For the standardized curve on the right, the units of measurement are just standard deviations.

3. Label the mean of the curve. Look back at your null hypothesis to label the mean of the normal or Student t curve. The mean of the standardized curve is always zero.

Label the Standard Deviation of the Curve

This is where many students get messed up. They don't realize that they need to do a calculation for the standard deviation and, if they do realize it, they don't realize that the calculation differs depending on what scenario they have. Below, we show the three possible curve pairs, along with their corresponding means and standard deviations. Notice that the standard deviation calculations are different for each:

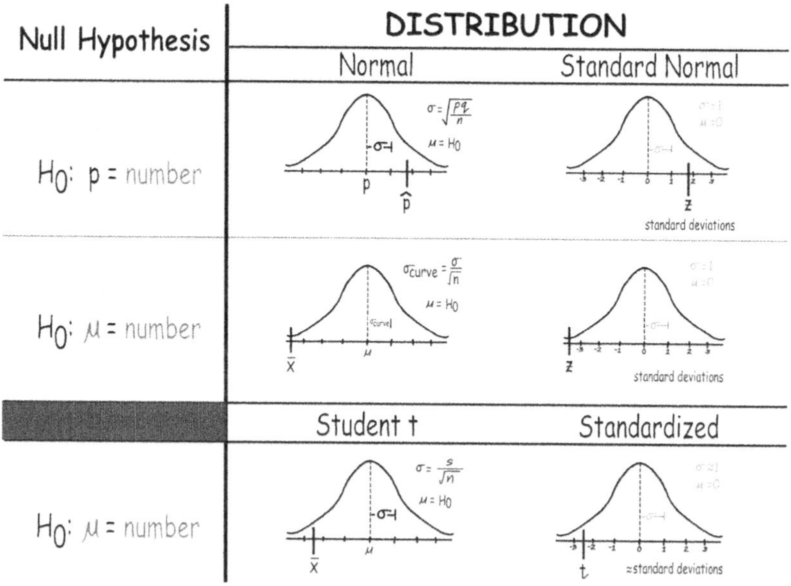

After calculating the standard deviation of the curve, you use it to fill out the values on the x-axis. The labeling of the axis begins in the center of the curve at the mean. Then, use the standard deviation you just calculated to label three standard deviations above and below the mean.

Number The x-axis Of The Curve

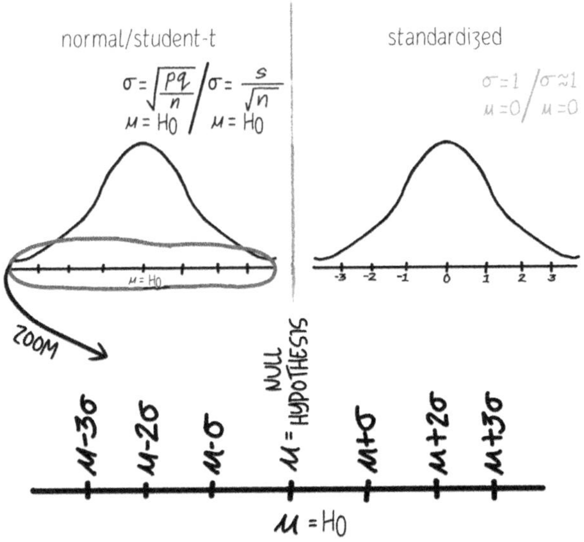

Shade the Significance Level Area

Next, we shade in the area of the significance level in the tail or tails of the curves. To do this, we use our alternative hypothesis and the significance level, **α**, to determine roughly how much of the tails of the graph to shade. The significance level and corresponding shaded area will be the same in both graphs. If needed, see the earlier section on significance levels for more details.

Label the Critical Value(s) and Rejection Region

Next, we compute the critical values, which are values on the x-axis that set the boundary for both the significance level area and the rejection region. This is probably the most important calculation and one that many students miss. How you calculate the critical values depends on what distribution you have and the number of tails in your test. If needed, see the section on critical values for examples of and calculator inputs for critical values.

Once you've worked through this checklist and labeled all the curves, you're ready to do your hypothesis test!

11. STEP 4 DO THE TEST

The goal of Step 4 is to do the hypothesis test. There are two tests to choose from: The P-Value Test or the Critical Value Test. At the end of this step, you will either reject the null hypothesis or fail to reject it. Step 4: Do the Test includes:

1. Calculate the p-value and test statistic.
2. Complete the curve sketch.
3. Do the P-Value Test.
4. Do the Critical Value Test.

We'll walk through the P-Value and Critical Value Tests separately, but here are the quick versions of each:

- The P-Value Test requires you to calculate the p-value and check the following criteria:

> P-value low, the null must go
> P-value $\leq \alpha \Rightarrow$ Reject H_0
>
> P-value high, the null will fly
> P-value $> \alpha \Rightarrow$ Fail to Reject H_0

- The Critical Value Test requires you to calculate the test statistic and check a slightly different criterion:

observation/test statistic
in the rejection region, towards the tail
⇒ Reject H_0

observation/test statistic
not in rejection region, closer to the mean
⇒ Fail to reject H_0

Let's learn what calculations we need and how to do them.

Calculate the P-value and Test Statistic

Before we jump into the calculations, let's define a few important terms:

- **P-value:** It's the probability of getting the value of the test statistic that is *at least as extreme* as the test statistic obtained from the sample, given that the null hypothesis is true. It is represented by an area under the curve to the left or right of the test statistic.
- **Observation:** A sample statistic is observed during a study and is called the observation. It is a sample statistic seen (observed) in the real world, and it has real-world units of measure. This observation will be placed on the normal or Student t distribution, and the x-axis will have units of measure.
- **Test Statistic:** The test statistic is what we get when converting the observation to a standard score. The test statistic goes on the x-axis of the standardized curve.

The image below shows the location of the p-value, the observation, and the test statistic:

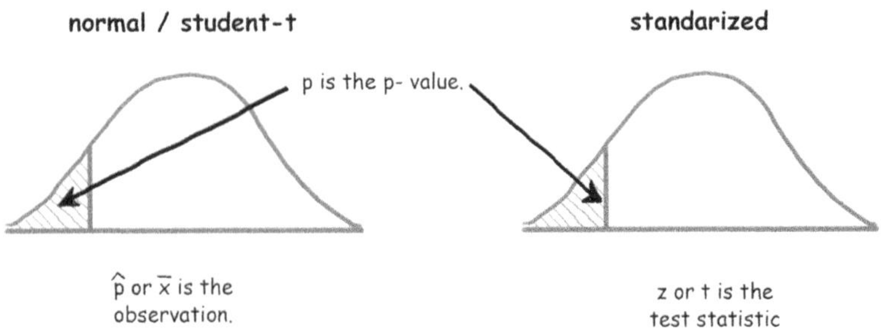

In the Foundation section, we covered three different ways to calculate p-values and test statistics:

1. **Use the TI-84 Test Functions:** For hypothesis tests about proportions, use 5:1-PropZTest. For hypothesis tests about a mean (with σ unknown), use 2:T-Test. In the unlikely event that you have a hypothesis test about a mean (with σ known), use 1:Z-Test.
2. **Use the Cumulative Distribution Function** (for p-values): To find a p-value for a hypothesis test about a proportion, use 2:normalcdf(). To find a p-value for a hypothesis test about a mean (with σ unknown), use 6:tcdf(). To find a p-value for a hypothesis test about a mean (with σ known), use 2:normalcdf().
3. **Use a Formula** (for the test statistic): Given the observation, you can find the test statistic (which is a standard score, i.e., z-score or t-score) by plugging in your observation, mean, and standard deviation into the standard score formula:

equation in words

$$\frac{\text{observation - mean}}{\text{standard deviation}} = \text{standard score}$$

Calculate the p-value (for the P-Value Test) or the test statistic (for the Critical Value Test) and fill in the following chart. Note that you also need to fill in the observation, which is a sample statistic given to you in the question.

Sketching Curves Checklist

	CHECKLIST	OBSERVATION SPACE	STANDARDIZED SPACE
1.	observation/ test statistic ☐ (x-axis)	observation $\hat{p}$ or $\bar{x}$ =	test statistic z or t =
2.	p-value ☐ (area)	same area, both curves p-value = ____	Left Tailed Right Tailed Two Tailed

Complete the Curve Sketch

Now, you have all the information you need to do the P-Value Test or the Critical Value Test. The thing that is different about the tests is that they compare different things. The P-Value Test compares areas, whereas the Critical Value Test compares values on the x-axis of the graph. Both tests, however, use the same graph and will give you the same answer.

Here's a sneak peek of the completed curve for the hypothesis test about a 1-sample population proportion:

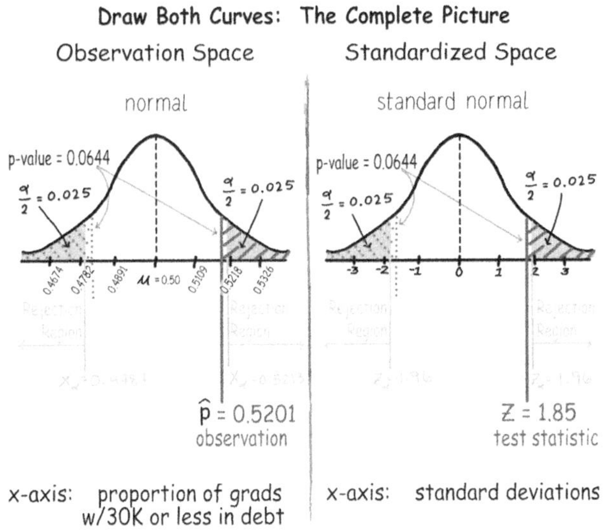

Steps 3 and 4 add a lot of information to the curves. Don't worry; we'll discuss this in Parts III and IV.

Do the P-value Test

Before we jump into the P-Value Test, here are some important facts to review about p-values:

1. A p-value (short for probability value) is an area under the curve.
2. It is the same size on both the normal and standard normal curves; the shaded areas are the same.
3. More specifically, it is the area under the curve in the tail of the observation or test statistic.
4. It represents the probability of getting the value of the test statistic that is *at least as extreme* as the test statistic obtained from the sample, given that the null hypothesis is true.

5. In the P-value Test, you compare the p-value to the significance level area, α.

To make a decision, follow the rule for the P-Value Test:

P-value low, the null must go
P-value $\leq \alpha \Rightarrow$ Reject H_0

P-value high, the null will fly
P-value $> \alpha \Rightarrow$ Fail to Reject H_0

Many students try to memorize this blindly, and sometimes they get lucky and get the correct answer, but you'll always get the right answer if you understand what the rule means. To do that, let's look at how this rule is visually represented in a graph.

The following graph shows two different scenarios. The left column shows scenarios where the p-value (striped area) is *less* than the significance level area, α (shaded area), and, therefore, we reject the null hypothesis. The right column shows the scenarios where the p-value (striped area) is *greater* than the significance level area, α (shaded area), and, therefore, we *fail to reject* the null hypothesis.

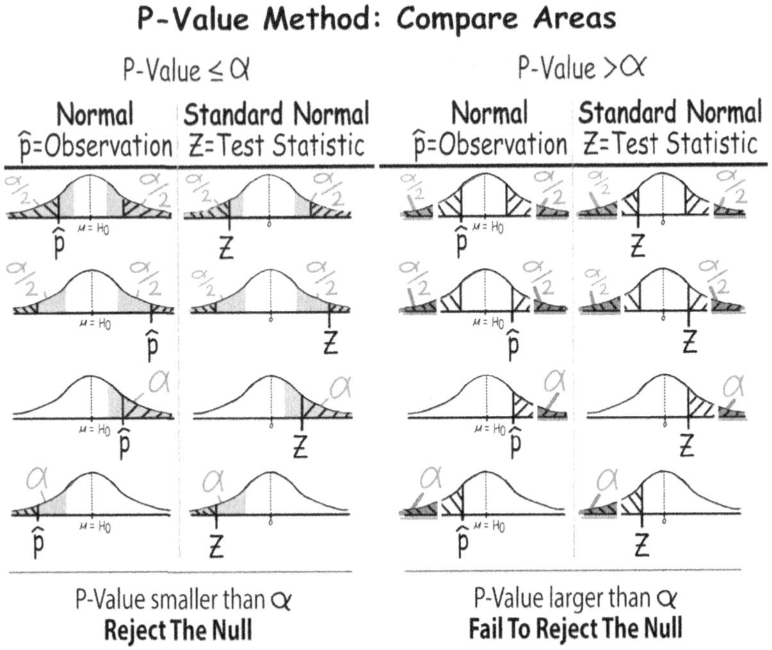

Notice that in each of the graphs, we are comparing the striped area (the p-value) to the shaded gray area (the significance level area, **α**). The graphs tell us if the p-value is larger or smaller than the significance level, **α**. Visually, this is how we determine if we'll reject or fail to reject the null hypothesis.

A+ Student Question: Why do we double or half the p-value in two-tailed tests, and how do we know which to do? Check out the answer, along with answers to other *A+ Student Questions,* at allisonlovesmath.com/hypothesis-testing.

Do the Critical Value Test

The Critical Value Test focuses on values on the x-axis of our curves that we drew in Step 3.

Before we get to this point, we should have already:

1. Labeled the critical value(s) and rejection region.

2. Found and drawn the observation and test statistic on the x-axis.

Now, we make a decision using the rule for the Critical Value Test:

> observation/test statistic
> in the rejection region, towards the tail
> ⇒ Reject H_0
>
> observation/test statistic
> not in rejection region, closer to the mean
> ⇒ Fail to reject H_0

Much like the P-value Test, many students try to memorize the Critical Value Test without understanding it. You're not the kind of student who likes to torture themselves, though. You're the kind of student who knows that understanding something is the faster, less stressful, and more worthwhile way to learn things.

To better understand the Critical Value Test, look at how the rule is visually represented in a graph. To do that, we've taken the same graphs that we looked at for the P-value Test and zoomed in on just the x-axis. Why? Because it'll help you to remember that the Critical Value Test compares values on the x-axis.

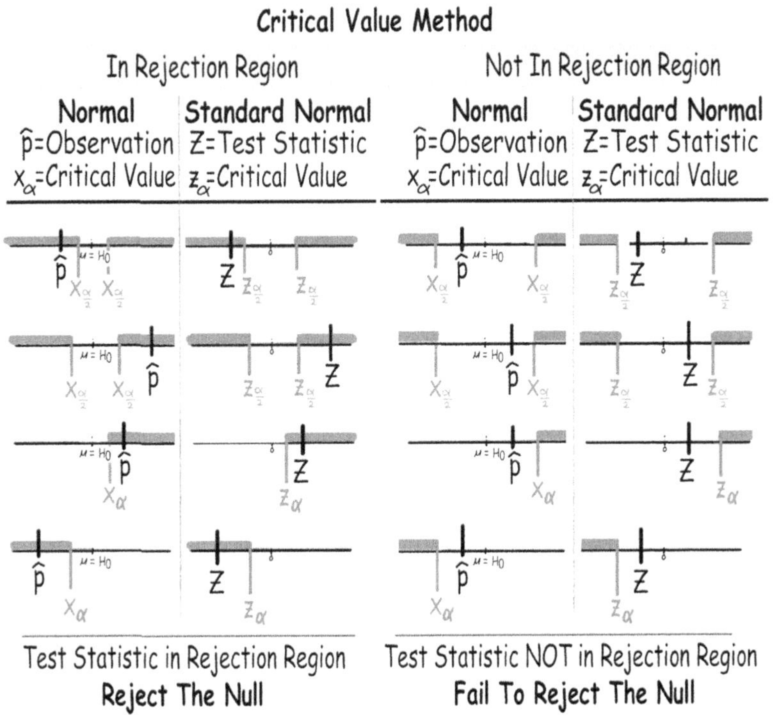

The image above shows two different scenarios. First, graphs on the left show scenarios where the observation and test statistic fall in the rejection region and, therefore, we reject the null hypothesis. In each of these scenarios, the observation and test statistic:

- are further from the mean than the critical value
- fall in the rejection region
- are between the tail and the critical value

Those are just three different ways of saying the same thing.

Second, let's look at the scenarios in the right column, where we fail to reject the null hypothesis. In each of these scenarios, the observation and test statistic:

- are between the mean and the critical value
- Are *not* in the rejection region
- Are not between the tail and the critical value

Summary of the P-Value and Critical Value Tests

The P-Value Test and the Critical Value Test yield the same results. If the p-value is smaller than the significance level area, we reject the null hypothesis. If the test statistic is contained in the rejection region, we reject the null hypothesis. You will get the same result regardless of whether you're using the P-Value Test or the Critical Value Test.

To make showing your work easier for you, we've created a flow chart to help you through both tests:

STEP 4: DO THE TEST

TEST	P-VALUE METHOD		CRITICAL VALUE METHOD	
CALCULATE	P-Value =		Test Statistic =	
TEST	Is P-Value $\leq \alpha$?		Is Test Statistic in the Rejection Region?	
	No ↓	Yes ↓	No ↓	Yes ↓
DECISION	Fail to Reject H_0	Reject H_0	Fail to Reject H_0	Reject H_0

12. STEP 5 STATE THE CONCLUSION

Now that you've completed your test and know if you've rejected or failed to reject the null hypothesis, you need to interpret what that means. This can be confusing for students because the wording is confusing. To correctly word your conclusion, first, follow the flow chart we created for you, then write the conclusion in your own words.

Follow the Flow Chart

To use the flow chart, you need two pieces of information from your hypothesis test:

- Your original claim, which is either your null or alternative hypothesis from Step 1: State the Hypotheses.
- Your decision, which is to reject or fail to reject the null hypothesis, from Step 4: Do the Test.

Circle these two pieces of information in the flowchart to get the opening to your conclusion.

STEP 5: STATE YOUR CONCLUSION

Step 1: Claim	Step 4: Result	Conclusion
Claim is H_A	→ Reject H_0 →	There is sufficient evidence to support the claim that....
	→ Fail to Reject H_0 →	There is not sufficient evidence to support the claim that....
Claim is H_0	→ Reject H_0 →	There is sufficient evidence to warrant the rejection of the claim that....
	→ Fail to Reject H_0 →	There is not sufficient evidence to warrant the rejection of the claim that....

For the first part of the sentence, copy the words directly from the flow chart. For the second part of the sentence, look back to Step 1 again, and plug in the wording from your claim. It seems simple enough, but many students really botch up their conclusion.

Common Mistake: Assuming your claim is the null hypothesis. Because Step 4: Do the Test is written in terms of the null, there is a desire to write a conclusion about the null. However, the null hypothesis is not necessarily the original claim. Here's how to avoid that mistake. Look back at your work in Step 1, and remember that: **H_0 *is not necessarily the claim.***

There are tons of examples where H_a is the claim!

Write the Basic Conclusion

The wording from the flow chart is intended to help you get started and use the basic language that your teacher or homework might be looking for. However, your job is not done until you can state the conclusion in plain English, in your own words.

When you write your conclusion, you should directly address the original claim from the problem statement; read the claim from Step 1, and ask yourself, "What was the question asking about? Was any additional context provided in the question?" Write a conclusion that is between one and three sentences. Your basic conclusion needs to tie together:

- the initial claim in Step 1,
- the conclusion of the test in Step 4,
- and the scripted conclusion produced by the flow chart.

Common Mistake: Not understanding that "failing to reject" H_0 is not the same as "accepting" or "proving" H_0. It's a slightly weaker statement, so replacing it with the words "accepting" or "proving" is incorrect. We're often told not to use double negatives when writing English papers, but Statisticians care more about the strength of the statement than they do about grammar.

Common Mistake: Accidentally writing the conclusion about a sample, not a population. No! Your hypothesis test and your conclusion are statements about a population, not a sample. It's obvious once you've got the hang of it, but easy to forget when so many different pieces of information are thrown at you when you're first learning hypothesis testing.

Common Mistake: Mistakenly writing the conclusion about the null hypothesis, instead of your claim. Your conclusion must address the original claim in Step 1, and in many questions, the alternative hypothesis is the claim! It's so simple but so easy to forget. To remember this, make sure you refer back to your claim in Step 1.

Write the Meaningful Conclusion

The last thing that some teachers do is ask you to write a conclusion in your own words. What does the conclusion mean? How does it affect someone's decisions? What are the consequences of the results? This is the part that proves you understand what the hypothesis test was about and why it was done in the first place.

Often, my students miss this because it's not required by our online homework tool. I'll emphasize it in class, but because it's not required in their online homework, they don't put much time into it and miss it on the exam. If your teacher emphasizes this last step in lectures or notes, be sure to practice it on homework because it'll likely show up on your exam.

PART III

HYPOTHESIS TESTING FOR A 1-SAMPLE POPULATION PROPORTION

This section guides you step-by-step through how to solve a 1-Sample hypothesis testing question about a population mean.

Step 1: State the Hypotheses will help you identify the keywords in the prompt that indicate that a question is asking about a population proportion (and not a population mean). We'll help you clarify what notation to use, find the complement, and correctly state your null and alternative hypotheses.

Step 2: Choose a Distribution walks you through how to know which parameter you're using, how to check requirements, and how to confidently determine which test to use.

Step 3: Sketch the Curves is the most frequently forgotten step, but also the most important. In this step, we'll show you how to graph the normal curve and the standard normal curve. We'll break down how to calculate and label values for the x-axis. Then, we show you how to shade in the significance level area, $\boldsymbol{\alpha}$, and identify the rejection region interval on the x-axis.

Step 4: Do the Test shows you how to pick out the observed sample statistic in the problem statement and transform it into a p-value or test statistic. Then we complete the curve sketch in the observation space and the standardized space. Lastly, we follow the P-Value Test or Critical Value Test to carry out the test and make a decision about the null hypothesis.

Step 5: State the Conclusion provides an easy-to-use flow-chart to help you figure out what wording to use in your final conclusion, and prepare you for how to write a more meaningful conclusion should your teacher require one.

13. THE 1-SAMPLE POPULATION PROPORTION QUESTION

Here's the hypothesis testing question for a population proportion that we'll be discussing in detail in the pages ahead:

> A newspaper article published that, in the United States, fifty percent of college students graduate with $30,000 or less in student loans. A college wants to know if their recent graduates differ from the national population; they claim that their graduates are different. They surveyed a simple random sample of 2,113 of their recent graduates. They found that 1,099 of them had a total of $30,000 or less in loan debt. Test the claim at a 0.05 significance level.

Now, there is a ton of information packed into these six sentences. Many hypothesis testing questions contain more information than the word problems you've seen in your Algebra classes. If you're feeling intimidated by it, you're not alone!

But does that intimidation mean we give up? Absolutely not! It just means we have to remind ourselves that learning to do difficult things is worthwhile.

Here are two suggestions for conquering this intimidating question:

1. **Try to solve it yourself first.** Active learning,

where you try it yourself first, then review the answers to learn from your mistakes, is more effective than passive learning, where you read through the answers without doing the work yourself.

2. **Break it down into bite-sized pieces.** If you can't do the entire problem, at least try each step and compare your outcome to the solution here. Once your outcome from a step matches, move on to the next one. If you practice, you'll soon be able to work an entire problem, just checking the final solution at the end.

Active learning plus breaking it into small steps is the path to success for anything difficult, including hypothesis testing questions.

Ready? It doesn't matter. You might never feel ready, so give it a shot and try to get through the question on your own. Then, check your work on the next page.

Solution

Woohoo! You tried the question yourself first! Whether you get it correct or not, give yourself a pat on the back for trying. That is your first huge step to success.

Below, we break down the five steps for you. Read through, check your work, and write down any mistakes or questions you have.

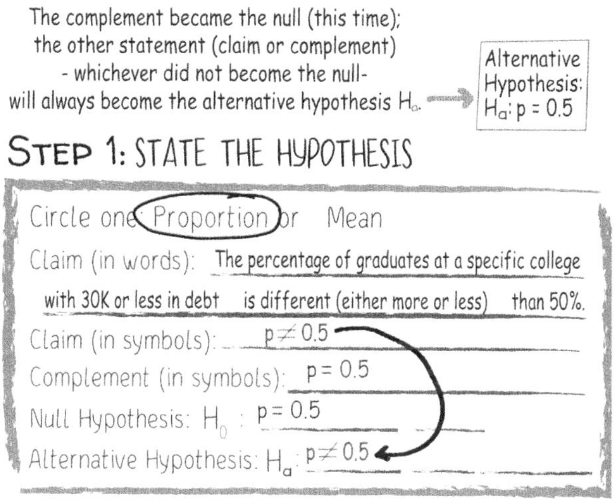

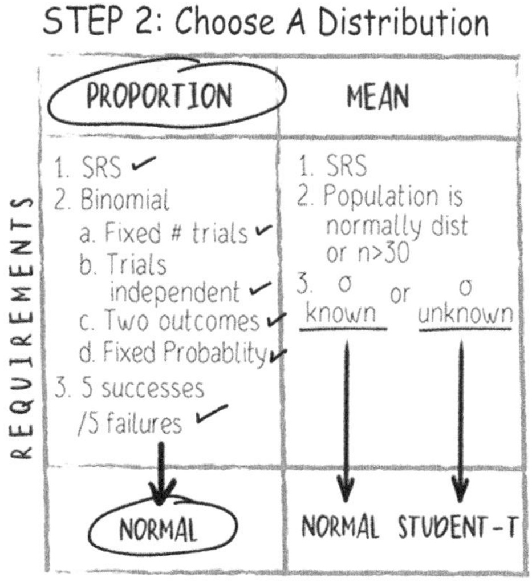

For Step 3: Sketch the Curves, fill in the following checklist as you track what you need to label on the curves.

Sketching Curves Checklist

	CHECKLIST	OBSERVATION SPACE	STANDARDIZED SPACE
1.	distribution ☑	normal / Student t	standard normal / standardized Student t
2.	units (x-axis) ☑	proportion of grads w/30k or less debt	NO CHOICE standard deviations
3.	mean of curve ☑	μ = Null Hypothesis p = 0.50	NO CHOICE $\mu = 0$
4.	standard deviation of curve ☑	$\sqrt{\frac{pq}{n}} = \sqrt{\frac{(.5)(.5)}{2113}}$ ≈ 0.01088	circle one 1 or $\approx$ 1
5.	significance level α (area) ☑	same area, both curves α = 0.05	Left Tailed Right Tailed Two Tailed
6.	critical value (x-axis) ☑	$X_{\alpha/2}$ left = 0.4786 right = 0.5213	$Z_{\alpha/2}$ $-Z_{\alpha/2} = -1.96$ $Z_{\alpha/2} = 1.96$

Here are the fully filled out graphs for Step 3: Sketch the Curve:

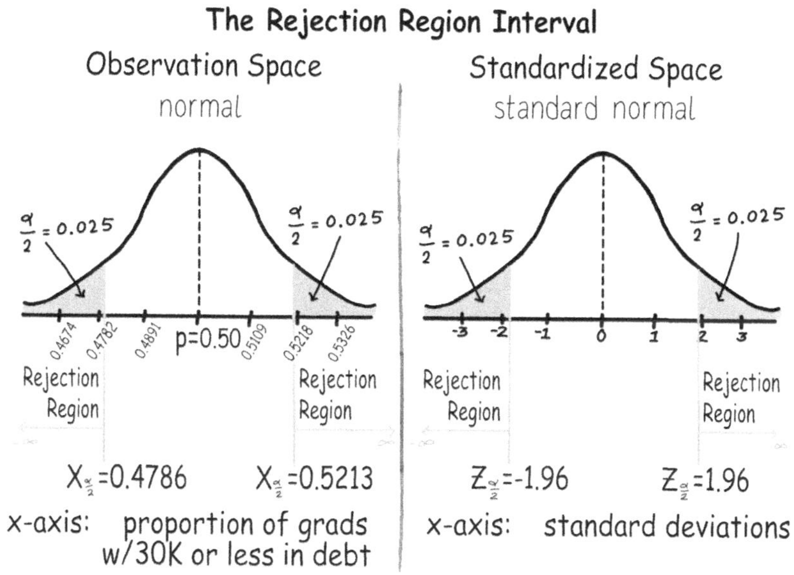

Here are the final calculations, graphs, and decisions for Step 4: Do the Test:

Sketching Curves Checklist

	CHECKLIST		OBSERVATION SPACE	STANDARDIZED SPACE
1.	distribution	☑	**normal** / Student t	**standard normal** / standardized Student t
2.	units (x-axis)	☑	proportion of grads w/30k or less debt	NO CHOICE standard deviations
3.	mean of curve	☑	μ = Null Hypothesis $p = 0.50$	NO CHOICE $\mu = 0$
4.	standard deviation of curve	☑	$\sqrt{\frac{pq}{n}} = \sqrt{\frac{(.5)(.5)}{2113}}$ ≈ 0.01088	circle one **1** or ≈ 1
5.	significance level α (area)	☑	same area, both curves $\alpha = 0.05$	Left Tailed Right Tailed **Two Tailed**
6.	critical value (x-axis)	☑	$X_{\alpha/2}$ left = 0.4786 right = 0.5213	$Z_{\alpha/2}$ $-Z_{\alpha/2} = -1.96$ $Z_{\alpha/2} = 1.96$
1.	observation/ test statistic (x-axis)	☑	observation $\hat{p} = \frac{\#\ successes}{n} = \frac{1099}{2113}$	test statistic $Z = 1.85$
2.	p-value (area)	☑	same area, both curves p-value $= 2*0.0322$ $= 0.0644$	Left Tailed Right Tailed **Two Tailed**

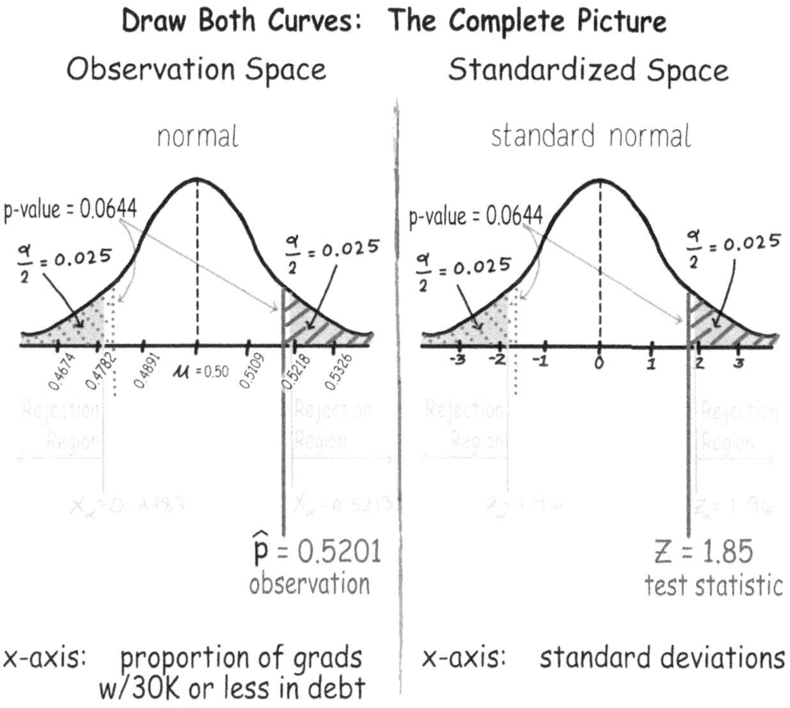

STEP 4: DO THE TEST

TEST	P-VALUE METHOD		CRITICAL VALUE METHOD	
CALCULATE	P-Value = 2*0.0322 = 0.0644		Test Statistic = 1.85	
TEST	Is P-Value ≤ α ? p-value = 0.0644 > α = 0.05		Is Test Statistic in the Rejection Region?	
	(No)	Yes	(No)	Yes
DECISION	(Fail to Reject H_0)	Reject H_0	(Fail to Reject H_0)	Reject H_0

For Step 5: State the Conclusion, here is the flow chart and your conclusion:

STEP 5: STATE YOUR CONCLUSION

Step 1: Claim	Step 4: Result	Conclusion
Claim is H_A	→ Reject H_0 →	There is sufficient evidence to support the claim that....
	→ Fail to Reject H_0 →	There is not sufficient evidence to support the claim that....
Claim is H_0	→ Reject H_0 →	There is sufficient evidence to warrant the rejection of the claim that....
	→ Fail to Reject H_0 →	There is not sufficient evidence to warrant the rejection of the claim that....

Basic Conclusion: There is not sufficient evidence to support the claim that the percentage of graduates at a specific college with $30,000 or less debt is different than 50%.

How'd you do? If you got any of the steps wrong or didn't understand why you got your answers, don't stress! We're here to learn, and we do that best by trying it out and learning from our mistakes. Read the pages ahead to discover your mistakes, what the correct answers were, and how to avoid those mistakes in the future.

If you got the answer correct, great! Jump to the sections where you want to know more. If there is something you got correct, but you want to understand why or gain a deeper understanding of the process, jump to that section. We don't just learn from mistakes. We learn anytime we ask questions and we gain confidence anytime level up our understanding of a topic.

Let's dig into the details to understand exactly why and how to get the correct answers on your hypothesis testing homework, quiz, and exam questions.

14. STEP 1 STATE THE HYPOTHESES

The Step 1 building blocks:

- State the parameter.
- State the claim and complement.
- Write the hypotheses.

Step 1 requires some deliberation before stating the null and alternative hypotheses. These in-between blocks help you to figure out what the hypotheses should be.

While other books and study guides may skip over these basics, we don't because we've seen *many* students get confused here. They don't know how to identify keywords and important information in a word problem, so we'll show you exactly what to highlight and circle.

Step 1 is the complete story of how to identify the precise words that mold the claim and the complement and then how those two statements are transformed into the null and alternative hypotheses.

Let's walk through this step to translate the word problem into the null and alternative hypotheses.

State the Parameter

Read the problem statement and look for keywords; determine if the question is about a population mean or a population proportion.

> A newspaper article published that, in the United States, fifty percent of college students graduate with $30,000 or less in student loans. A college wants to know if their recent graduates differ from the national population; they claim that their graduates are different. They surveyed a simple random sample of 2,113 of their recent graduates. They found that 1,099 of them had a total of $30,000 or less in loan debt. Test the claim at a 0.05 significance level.

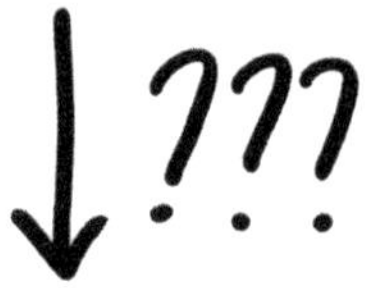

STEP 1: STATE THE HYPOTHESIS

Circle one: Proportion or Mean

Claim (in words): ______________________

____

Claim (in symbols): ______________________

Complement (in symbols): ______________________

Null Hypothesis: H_0 : ______ ______

Alternative Hypothesis: H_a ______ ______ ______

A mean is often indicated by the inclusion of sample data that has a mean and sample standard deviation; a proportion is often indicated by a percentage or ratio. Highlight keywords or phrases in the problem statement that identify sample statistics or a percentage.

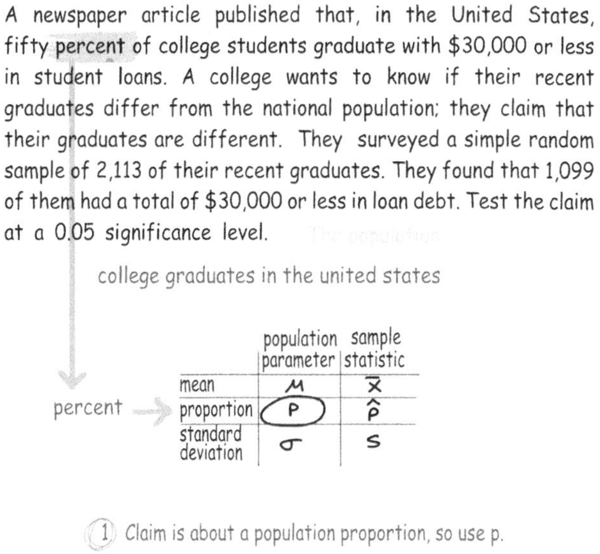

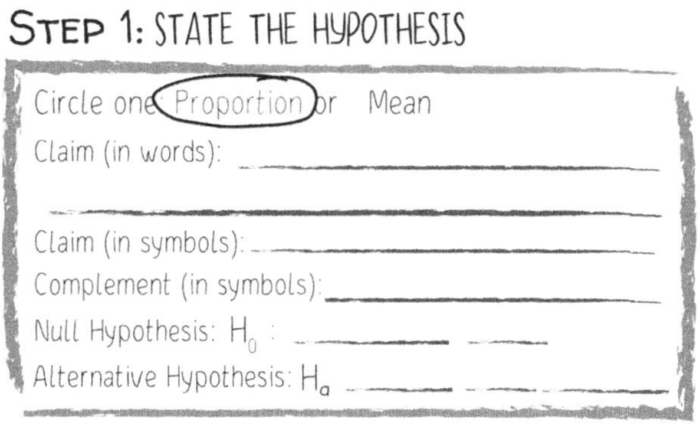

Pay special attention to the wording. The percentage describes all of the students in the United States who graduate with $30,000 or less in debt; this is our population. The sample, the subset of students in the United States from which we have data, is the group of graduates from a specific college with $30,000 or less in debt.

The question contains data from a sample proportion and asks a question about a population proportion; the claim will be written with the population proportion parameter, p.

State the Claim and Complement

Now that the parameter has been identified, you will express the claim in two different ways—first with words, then again with symbols.

A newspaper article published that, in the United States, fifty percent of college students graduate with $30,000 or less in student loans. A college wants to know if their recent graduates **differ from** the national population; they claim that their graduates are different. They surveyed a simple random sample of 2,113 of their recent graduates. They found that 1,099 of them had a total of $30,000 or less in loan debt. Test the claim at a 0.05 significance level.

(1) Claim is about a population proportion, so use parameter p.
(2) Rewrite 50% as a proportion, 0.50.
(3) "Differ from" implies larger or smaller, use $\neq$.

Put (1) (2) and (3) together: $p \neq 0.5$

STEP 1: STATE THE HYPOTHESIS

Circle one: (Proportion) or Mean

Claim (in words): The percentage of graduates at a specific college with 30K or less in debt is different (either more or less) than 50%.

Claim (in symbols): $p \neq 0.5$

Complement (in symbols): ______

Null Hypothesis: H_0 : ______

Alternative Hypothesis: H_a ______

After the claim is written in words, write it symbolically. Always write the claim using the population parameter; in this case, use p.

Make-It-Easier Tip: Always circle the keyword and highlight the claim when you first read through a hypothesis testing question.

Common Mistake: Plugging the claim in for the null hypothesis. It's important to remember that the null hypothesis is always an equality statement, meaning it must have an equal sign in the statement. In some cases, the claim will become the null hypothesis; in other cases, the claim will become the alternative hypothesis.

All Possible Claim Statements

Claim about a proportion	Claim about a mean
$p =$ number	$\mu =$ number
$p \leq$ number	$\mu \leq$ number
$p \geq$ number	$\mu \geq$ number
$p \neq$ number	$\mu \neq$ number
$p <$ number	$\mu <$ number
$p >$ number	$\mu >$ number

Next, write out the complement of the claim. By definition, a complement is the opposite of the claim. If the claim turns out to be false, then the opposite (i.e., the complement) must be true.

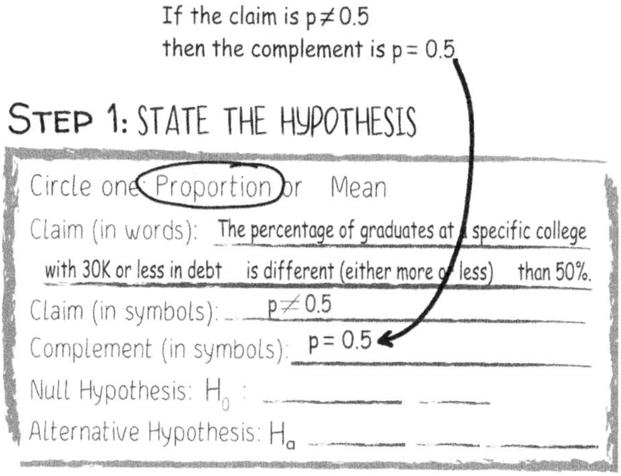

Write the Hypotheses

The third block is to write the hypotheses; state the null and alternative hypotheses. Look at both the claim and complement. The statement that contains equality will become the null hypothesis. The other statement will become the alternative hypothesis. In this example, the complement contains equality; in fact, it contains a strict equality. The complement will become the null hypothesis.

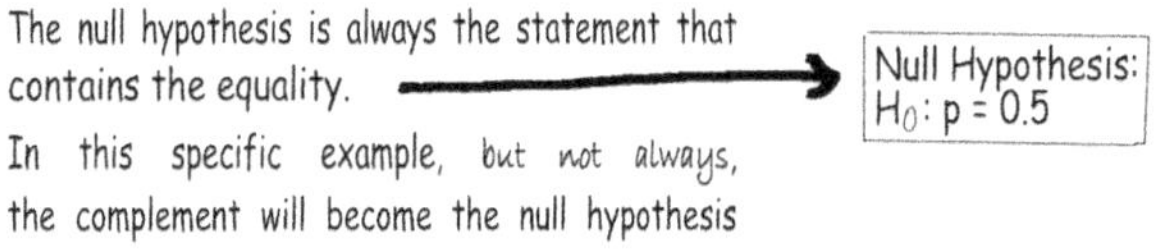

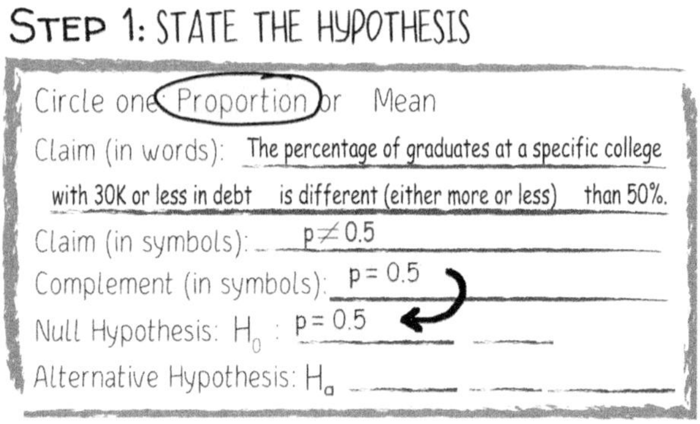

> It's tempting to plug in our claim as the null hypothesis, but that would be incorrect!

Common Mistake: Thinking our claim is the null hypothesis. In this question, it isn't! If our claim is an inequality (meaning our claim has > or < or ≠ in it), then our claim must be the alternative hypothesis. Why? Because the null hypothesis must have an equality in it. Always! That's the rule.

In the problem above, our claim is $p \neq .5$, so our claim will become the alternative hypothesis.

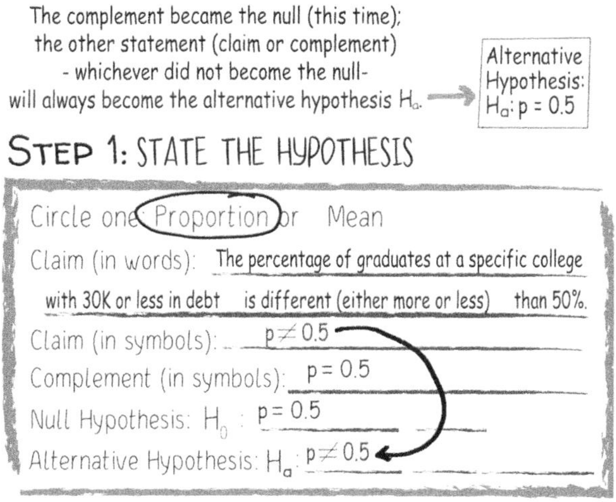

Let me reiterate one last time: The claim and the null hypothesis are not the same in this question. Our claim is the alternative hypothesis because it contains an inequality. The null hypothesis must always be an equality statement.

Why is this so important? Because if we botch up the null and alternative hypotheses, it'll botch up our conclusions at the end.

15. STEP 2 CHOOSE A DISTRIBUTION

These are the Step 2 building blocks:

- Restate the parameter.
- Check the requirements.
- Choose a distribution.

Since we are only looking at two types of hypothesis tests (i.e., population proportions and population means), there are only two options: the normal distribution or the Student t distribution.

Here is Step 2 already filled out so you can check your work.

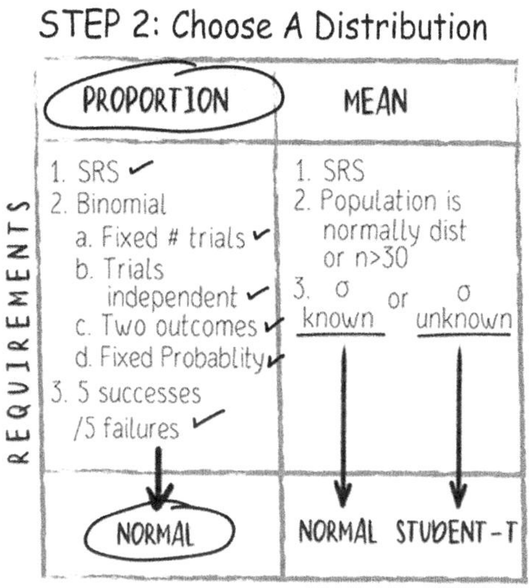

Read through the rest of the section to correct any mistakes or better understand why you got the answer you did.

Restate the Parameter

The first thing Step 2 asks is, "Is our test about a population proportion or a population mean?" This is important because:

- If the test is about a population proportion, we use the normal distribution.
- If the test is about a population mean, we use the normal or Student t distribution.

If you've done Step 1 correctly, you know this specific problem is about a proportion.

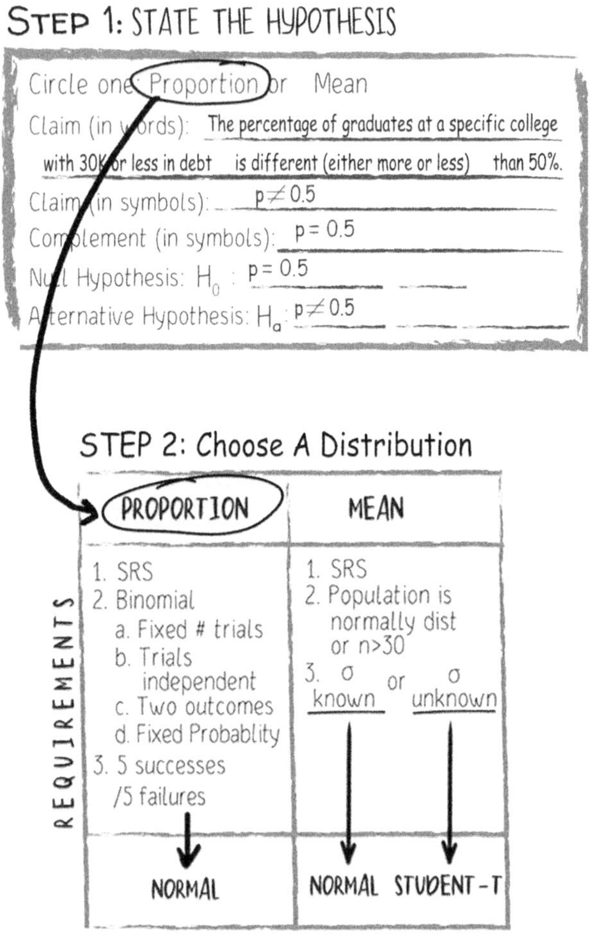

Done!

Check the Requirements

Don't fret if you're confused about how to check requirements. Hypothesis questions are notoriously bad about including all the information to show that requirements are met. Some questions state it obviously, yet others sneak in subtle wording, and others bypass the step entirely. It's a point of frustration for us as teachers and a point of frustration for our students as well. We, as teachers, tell our students to always prove that the requirements are met, then a homework question might leave necessary information out. It's a pain.

That being said, requirements are important. Don't, for a second, think you can go into the real world and do hypothesis tests without checking requirements!

The first requirement for a hypothesis test about a population proportion states that we need a simple random sample (i.e., SRS). Let's go back to the problem statement to see if any keywords tell us we have a simple random sample.

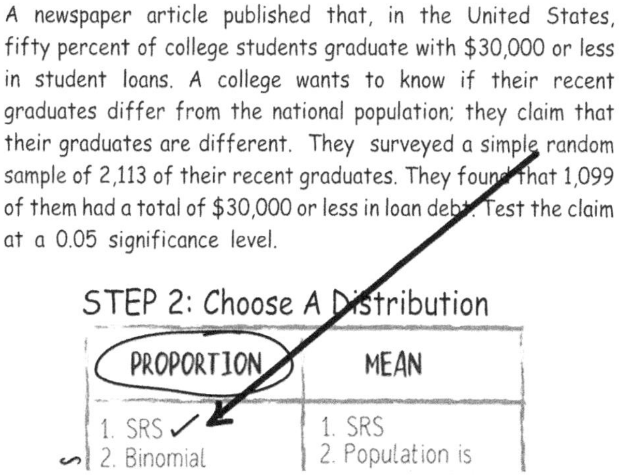

A newspaper article published that, in the United States, fifty percent of college students graduate with $30,000 or less in student loans. A college wants to know if their recent graduates differ from the national population; they claim that their graduates are different. They surveyed a simple random sample of 2,113 of their recent graduates. They found that 1,099 of them had a total of $30,000 or less in loan debt. Test the claim at a 0.05 significance level.

Second, we must check that the conditions of the binomial distribution are met:

1. Are there a fixed number of trials? Yes, the question states that they surveyed 2,113 recent graduates.
2. Are the trials independent? Yes. We assume this because we have a simple random sample.
3. Are there two outcomes? Yes. A graduate has either $30,000 or less in student loans or more than

$30,000 in student loans. There are no additional options.

4. Is the probability of success the same in all trials? Yes. It is reasonable to assume that the probability remains the same in all trials.

The image below highlights the wording in the problem statement that verifies (a) and (c).

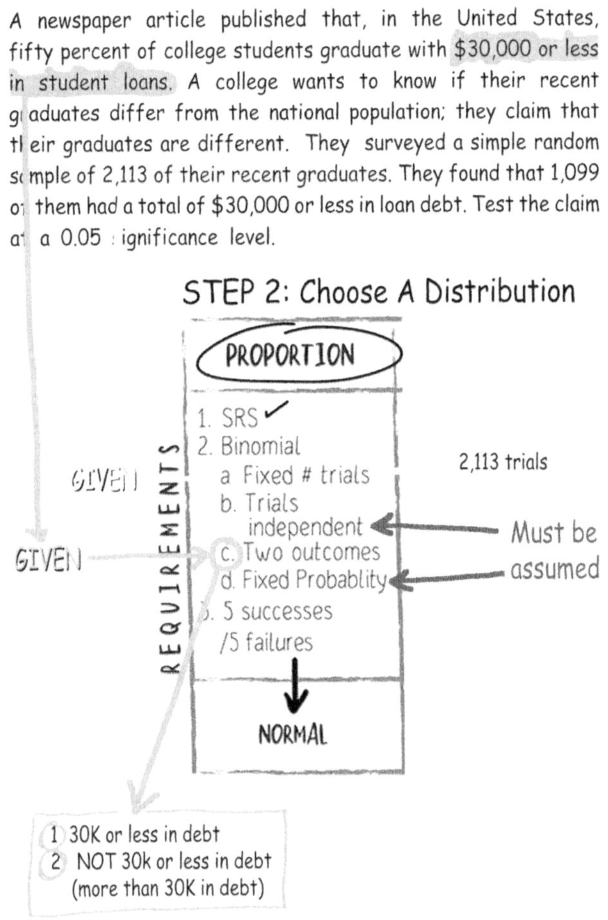

Third, we check that there are at least 5 successes and 5 failures. Again, look at the information in the question to verify that this is true.

> A newspaper article published that, in the United States, fifty percent of college students graduate with $30,000 or less in student loans. A college wants to know if their recent graduates differ from the national population; they claim that their graduates are different. They surveyed a simple random sample of 2,113 of their recent graduates. They found that 1,099 of them had a total of $30,000 or less in loan debt. Test the claim at a 0.05 significance level.

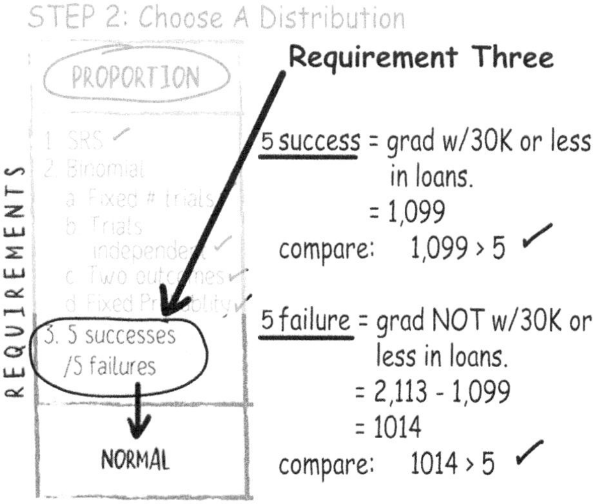

Different Wording: Another way to write 5 successes and 5 failures is:

- 5 successes = np where n is the number of trials and p is the assumed population proportion related to success.
- 5 failures = nq where n is the number of trials and q is the assumed population proportion related to failures, $q = 1\text{-}p$.

Some books ask you to verify that there are 5 successes and 5 failures. Others ask you to verify that $np \geq 5$ and $nq \geq 5$. These are simply different ways of asking the same thing.

Tough Class Tip: If your teacher emphasizes requirements, do not skip the requirements or simply say "meets requirements." If you have a teacher like us, you'll need to write out the requirements and state how you know each is met. Just because some teachers and texts skim over the requirements, they are extremely important in practice. If your teacher emphasizes them in class, then you should similarly emphasize them when showing your work.

Easy Class Tip: If your teacher and book don't emphasize the requirements, don't stress over them. There are many books, and consequently homework problems, that don't include enough information to verify the requirements. If this is your class and your teacher keeps saying things like, "just assume it meets the requirements," then follow your teacher's lead and feel free to breeze through them. Keep in mind, though, that, in practice, you'll need to be vigilant about requirements.

Choose a Distribution

Once the requirements for a population proportion hypothesis test are satisfied, follow the arrow in the chart! The arrow indicates the normal distribution is the appropriate choice for the hypothesis test.

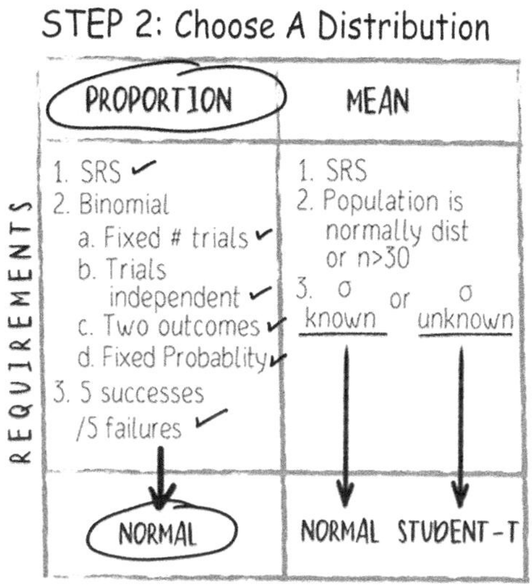

The hypothesis test performed will be the 1-sample test of proportions on the normal distribution.

Common Question: Why do we check requirements for a binomial distribution, when we are using the normal distribution for the hypothesis test? Great question! Under certain circumstances, we can use the normal distribution to approximate the binomial distribution. What circumstances? The exact circumstances we just checked: when we have an SRS, conditions for binomial distribution are met, and there are at least 5 successes and 5 failures.

16. STEP 3 SKETCH THE CURVES

At this point, most students want to jump directly to the P-Value Test or the Critical Value Test. Resist the urge! If you complete Step 3: Sketch the Curves, you will better understand the question and meaning of the results, as well as catch mistakes.

Sketching the curves breaks down into six things you need to label on the graph:

1. The distribution
2. The x-axis units
3. The mean of the curve
4. The standard deviation of the curve
5. The significance level area
6. The critical value(s) and rejection region

Here's the checklist that we'll fill in for our hypothesis test about a population proportion:

Sketching Curves Checklist

	CHECKLIST	OBSERVATION SPACE	STANDARDIZED SPACE
1.	distribution ☐	normal / Student t	standard normal / standardized Student t
2.	units (x-axis) ☐		NO CHOICE standard deviations
3.	mean of curve ☐	μ = Null Hypothesis	NO CHOICE $\mu = 0$
4.	standard deviation of curve ☐	circle one $\sqrt{\frac{pq}{n}}, \frac{\sigma}{\sqrt{n}}, \frac{s}{\sqrt{n}}$	circle one 1 or ≈ 1
5.	significance level α (area) ☐	same area, both curves α = ____	Left Tailed Right Tailed Two Tailed
6.	critical value (x-axis) ☐	$X_{\alpha}, X_{\alpha/2}$	$Z_{\alpha}, Z_{\alpha/2}, t_{\alpha}, t_{\alpha/2}$
	observation/test statistic (x-axis) ☐	observation $\hat{p}$ or $\bar{x}$ =	test statistic Z or t =
	p-value (area) ☐	same area, both curves p-value = ____	Left Tailed Right Tailed Two Tailed

Common Mistake: Not drawing the curves. The #1 way we've seen students raise their Statistics grades is by having them draw the curves. Why? Because so often, students who "kinda" understand hypothesis testing are able to state the hypotheses (Step 1) and can plug things into their calculator (Step 4) but don't know when they've made mistakes. Did you plug the wrong numbers into the calculator? Did you use the wrong calculator function? If you draw the curve, you can better understand when you get an incorrect answer. Because each Hypothesis Test question is worth many points on any given test, catching these mistakes is crucial to earning a strong grade in Statistics.

Therefore . . .

Graph the curves. The extra work will be annoying at first, *but* when you find yourself catching mistakes and getting more correct answers, you'll be thankful you added this extra step.

If you haven't already, take a moment to work through the checklist and draw curves for this problem yourself, then check your work at the end of Step 3. If your curves don't match, read through to find and learn from your mistakes.

Label the Distribution and Units

Labeling the distribution is a quick step. We look back at Step 2: Choose a distribution to see that because we have a hypothesis test about a population proportion, we will use the normal curve and the standard normal curve.

Step 3	Observation Space	Standardized Space
Distribution	normal (circled) / student-t	standardized
x-axis units/ label	Proportion of grads w/30K or less debt	NO CHOICE standard deviations
mean of curve	μ = Null Hypothesis $P = 0.50$	NO CHOICE $\mu = 0$
standard deviation of curve	$\sqrt{\frac{pq}{n}} = \sqrt{\frac{(.5)(1-.5)}{2113}}$ ≈ 0.01088	circle one: 1 (circled) or ≈ 1

Next, the units for the x-axis will be the proportion, p, of the population. In this question, it's the proportion of grads who graduate with \$30,000 or less student debt.

Step 3	Observation Space	Standardized Space
Distribution	normal/student-t (normal circled)	standardized
x-axis units/ label	Proportion of grads w/30K or less debt	NO CHOICE standard deviations
mean of curve	μ = Null Hypothesis $P = 0.50$	NO CHOICE $\mu = 0$
standard deviation of curve	$\sqrt{\frac{pq}{n}} = \sqrt{\frac{(.5)(1-.5)}{2113}}$ ≈ 0.01088	circle one (1) or ≈ 1 (1 circled)

Label the Mean of the Curve

Next, we label the mean of the curve. The mean of the curve is the null hypothesis from Step 1. (In this example, we obtained the null hypothesis: p = 0.50 of students who graduate with $30,000 or less student debt.)

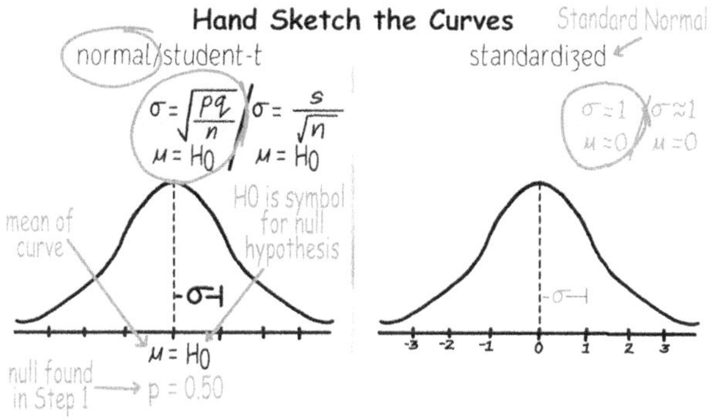

Step 3	Observation Space	Standardized Space
Distribution	normal/student-t (normal circled)	standardized
x-axis units/ label	Proportion of grads w/30K or less debt	NO CHOICE standard deviations
mean of curve	μ = Null Hypothesis $P = 0.50$	NO CHOICE $\mu = 0$
standard deviation of curve	$\sqrt{\frac{pq}{n}} = \sqrt{\frac{(.5)(1-.5)}{2113}}$ ≈ 0.01088	circle one: 1 (circled) or ≈ 1

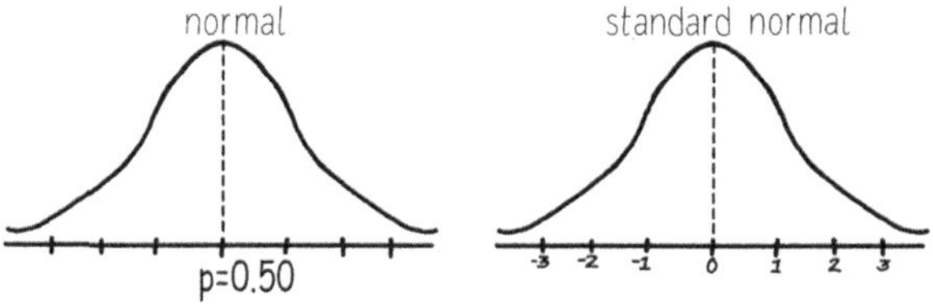

Common Question: For a hypothesis test about a population proportion, why is the mean p and not μ? For hypothesis tests, the mean is the null hypothesis, so for hypothesis tests about population proportions, the mean is the parameter p.

Label the Standard Deviation of the Curve

How you calculate your standard deviation depends on which distribution you are using and what population parameter your hypothesis test is about. In this example, our hypothesis test is about a population proportion and we are using the normal distribution, therefore, we use the following:

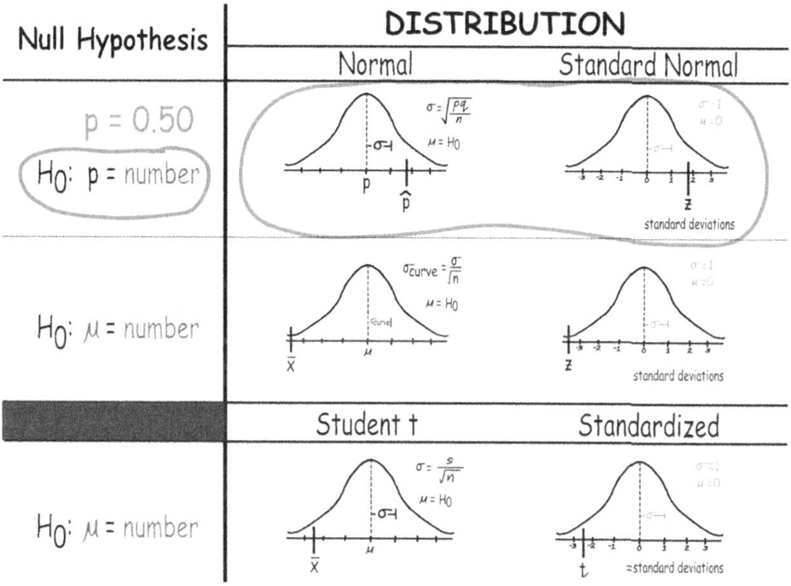

Plugging in values for the variables, we get:

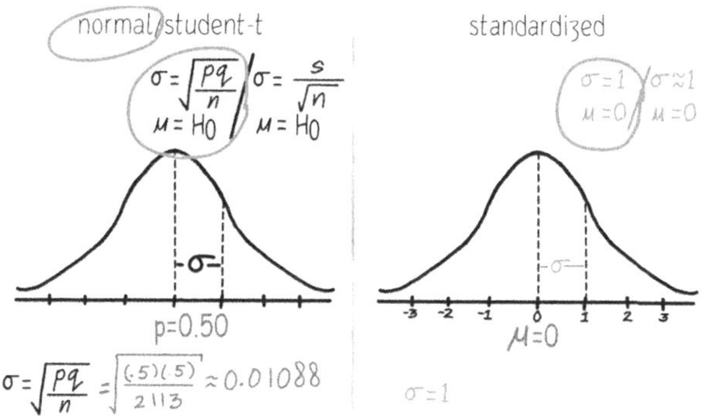

A+ Student Question: Where does the standard deviation of the curve formula come from? Great question! Many students just take their teacher's word for it and don't ask about this, but for students who are thinking about the *why* behind each step the equations for the standard deviation of the curve can feel like they were pulled out of thin air. If you want the details, check out the answer, along with answers to other *A+ Student Questions,* at allisonlovesmath.com/hypothesis-testing. (It's a bit long, and we don't want you getting side-tracked right here.)

Next, we'll use the mean and standard deviation to calculate our x-axis values, representing the number of standard deviations above and below the mean.

For the normal curve (on the left), we have $\mu = 0.5$ and $\sigma = 0.01088$. Here is how we calculate the values on the x-axis:

$$\mu - 3\sigma = 0.5 - 3 * 0.01088 = 0.4674$$
$$\mu - 2\sigma = 0.5 - 2 * 0.01088 = 0.4782$$
$$\mu - \sigma = 0.5 - 1 * 0.01088 = 0.4891$$
$$\mu = 0.5$$
$$\mu + \sigma = 0.5 + 1 * 0.01088 = 0.5109$$
$$\mu + 2\sigma = 0.5 + 2 * 0.01088 = 0.5218$$
$$\mu + 3\sigma = 0.5 + 3 * 0.01088 = 0.5326$$

Let's label these on the graph. Visually, it's hard to squish all the numbers into the graph, so we zoomed in on the x-axis for you:

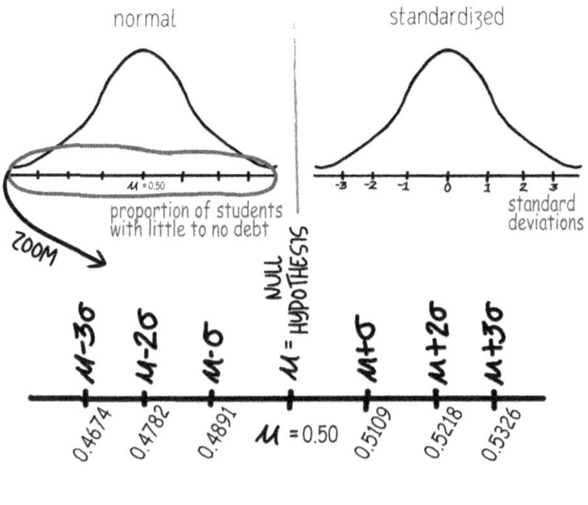

Common Question: How did we get the values on the x-axis of the standardized normal curve (on the right)? The x-axis of the standard normal curve is always the same. The mean is always zero, and the standard deviation is always one. This is why the x-axis for the standard normal is always labeled: -3, -2, -1, 0, 1, 2, 3.

Woohoo! Now that both x-axes are labeled, we can determine how many tails will be involved in the hypothesis test.

Shade the Significance Level Area

Our next step is to shade the area that corresponds to the significance level. The significance level area represents the probability of rejecting the null hypothesis when it is true. On the graph, where we shade α depends on whether we have a one-tailed or two-tailed test. We determine this by looking at our alternative hypothesis:

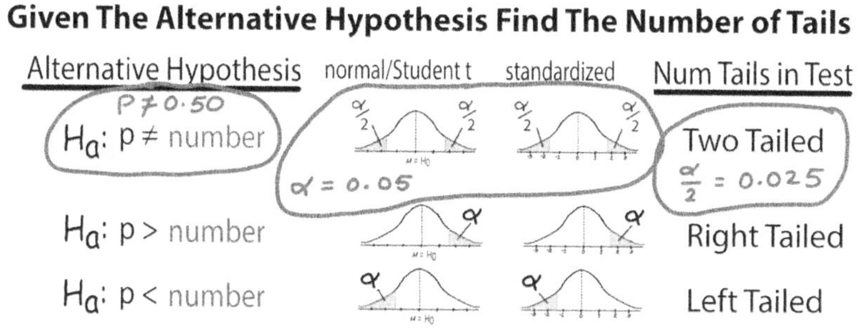

It's important to note that a two-tailed test is different from a one-tailed test. A two-tailed test tells us to split the significance level area between both tails and (in the next step) find two critical values. This means we'll shade $\alpha/2$ in both tails of the curve.

For our curves, we have:

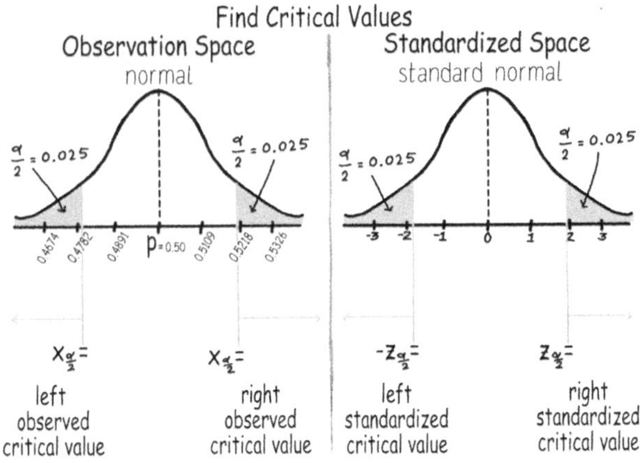

Notice that in the figure above, we shade *generally*. We don't know where the exact border to the significance level area starts until we find the critical values.

Label the Critical Values and Rejection Region.

Next, we need to find and label two critical values on both curves. Since this is a two-tailed test, we have four critical values to calculate.

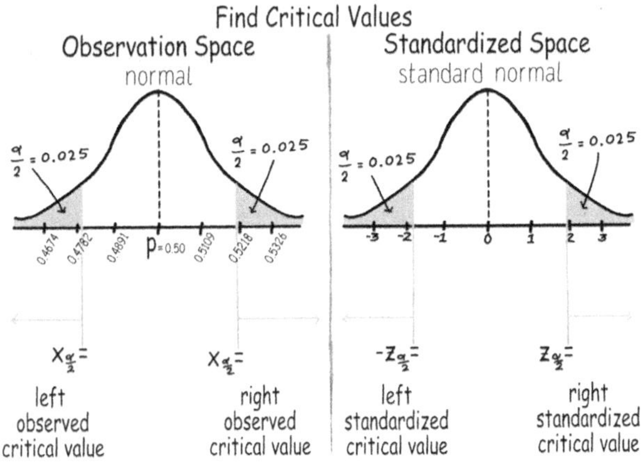

When calculating critical values on a TI-84 calculator, we use the invNorm(area, mean, standard deviation) function. There is one thing we must emphasize here:

- The invNorm(area, mean, standard deviation) function on your calculator requires your input to be the area *to the left* of the critical value.
- The total area under the curve is 1. Therefore, if **α**/2 is .025, then the area under the rest of the curve is 1-.025 = 0.975.

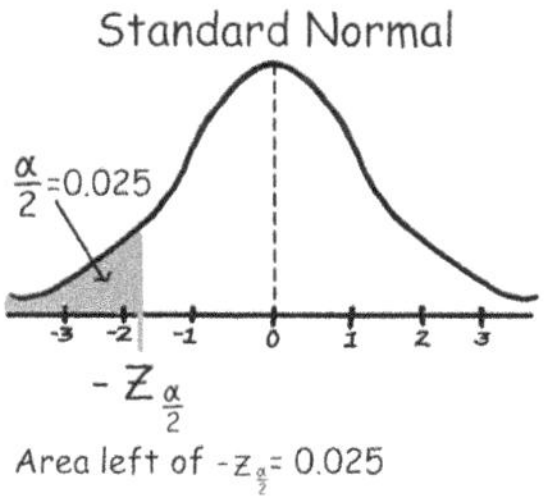

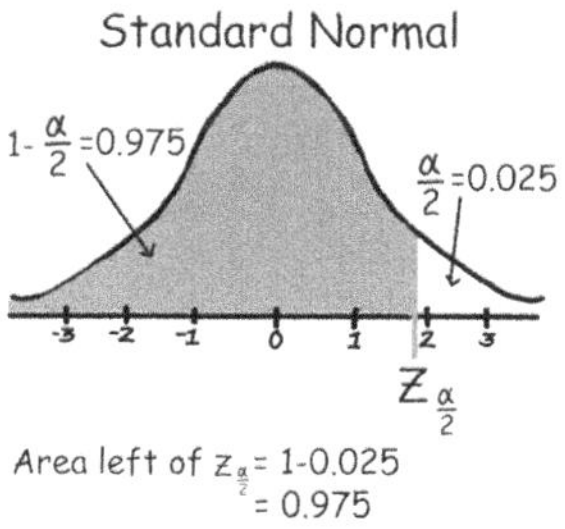

Let's calculate all four of our critical values.

Critical Value #1: Left

To find the standardized critical value to the left of the mean, use the function invNorm(0.025, 0, 1) = -1.959963986. When we label this critical value on our graph, we'll round the numeric output to two decimal places. Here's what the process looks like on your calculator and how that translates to your graph:

Output: $-Z_{\frac{\alpha}{2}}$, the left standardized critical value.

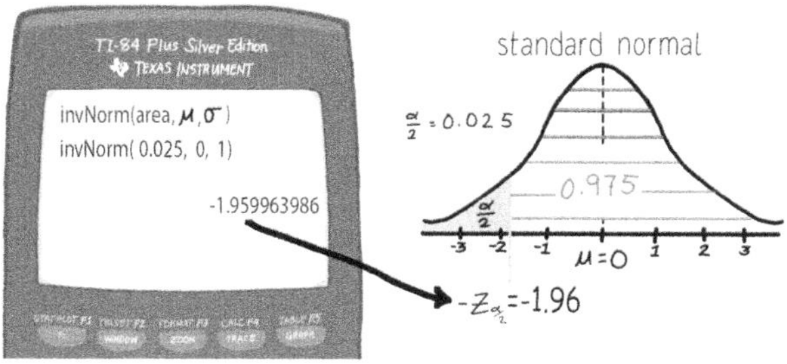

Critical Value #2: Right

To find the standardized critical value to the right of the mean, use the function invNorm(0.975, 0, 1) = 1.959963986. Why do we plug in 0.975 for the area? Because the invNorm() calculator function is programmed to use the area to the left of the critical value. Here's what the process looks like on your calculator and how that translates to your graph:

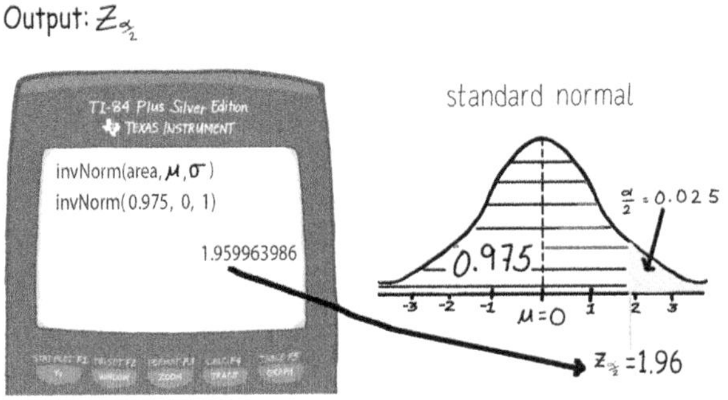

Notice anything interesting here? The two critical values are almost the same. The only difference is that one is negative and the other is positive! Why? Because the normal curve is symmetric about 0. Both areas in the tails are the same, so they'll be equidistant from the center.

Critical Value #3: Left

Next, let's find the critical values for the normal curve (on the left). Here, we use the mean and standard deviation of the curve that we previously calculated:

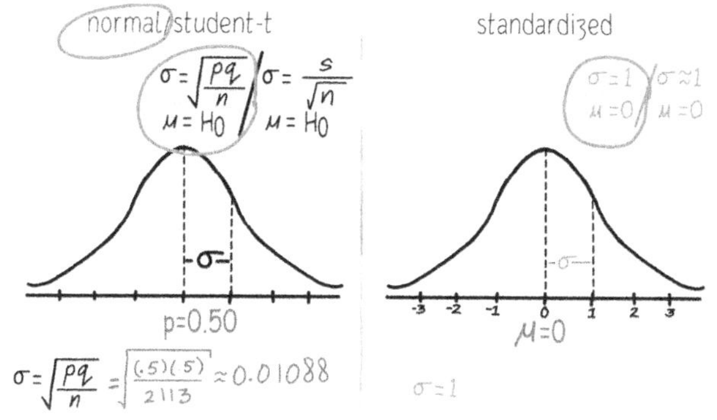

To calculate the left critical value on the normal curve, use the function invNorm(0.025, 0.5, 0.01088) = 0.4786755918. Here's what the process looks like on your calculator and how that translates to your graph:

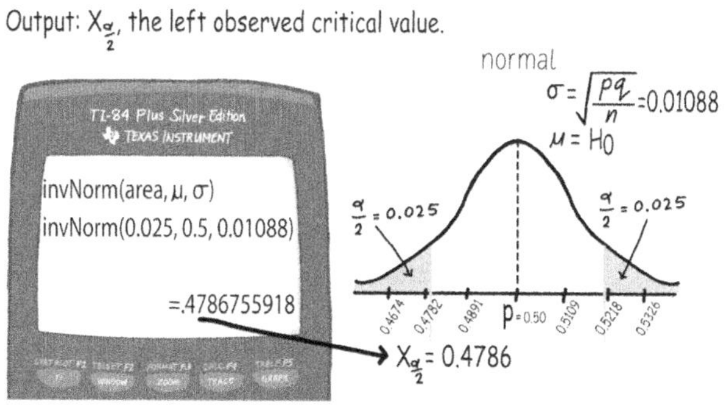

Critical Value #4: Right

To calculate the right critical value on the normal curve, use the function invNorm(0.975, 0.5, 0.01088) = 0.521344082. Remember to use 0.975 because we need the area to the left of the critical value. Here's what the process looks like on your calculator and how that translates to your graph:

Output: $X_{\frac{\alpha}{2}}$, the right observed critical value.

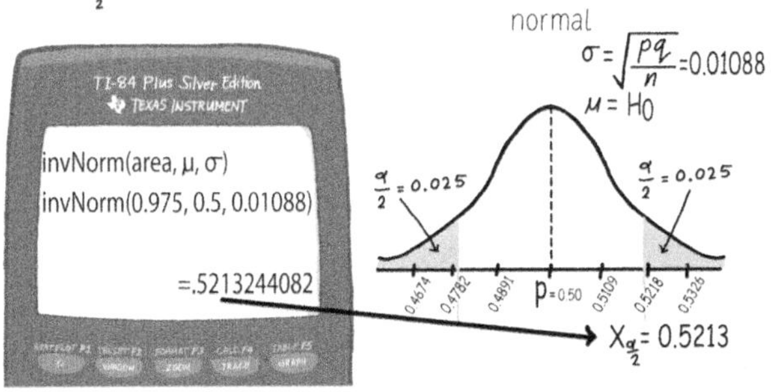

Now that we have all four critical values, what do we do with them? We use them to label the rejection region.

Label the rejection region.

A two-tailed test will have two rejection region intervals along the x-axis. One end of the interval is always at negative or positive infinity, and the other is at a critical value.

We have identified this interval along the x-axis by drawing an arrow originating at the critical value, pointing away from the mean. An observation or test statistic further away from the mean than the critical value is in the rejection region.

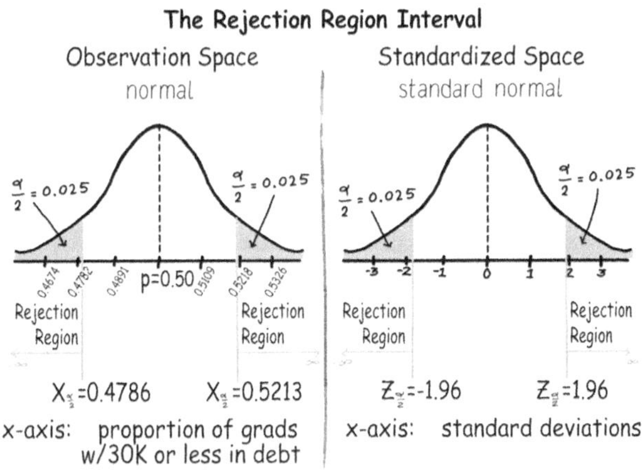

Some teachers and online homework tools like you to write out the rejection region as inequalities or in interval notation. If you need to do this, your rejection region intervals in the observation space (left graph) are:

- Inequality Notation: $x \leq 0.4786$ and $x \geq 0.5213$
- Interval Notation: $(-\infty, 0.4786]$ and $[0.5213, \infty)$

Your rejection region intervals in the standardized space (right graph) are:

- Inequality Notation: $x \leq -1.96$ and $x \geq 1.96$
- Interval Notation $(-\infty, -1.96]$ and $[1.96, \infty)$

Now that the axes, shaded areas, and critical values are all labeled, our curves are complete, and we are ready for Step 4!

A+ Student Questions:

- Why is the mean and standard deviation *of data* different from the mean and standard deviation *of the curve*?
- If I'm not allowed to use a calculator, how do I find critical values on the normal curve by hand?

Check out the answer, along with answers to other *A+ Student Questions*, at allisonlovesmath.com/hypothesis-testing.

17. STEP 4 DO THE TEST

Woohoo! We're through the first three steps. We've (1) stated the hypotheses, (2) checked which distribution to use, (3) drawn the curves, and set ourselves up for success here in Step 4: Do the Test. Which test? The P-Value Test or the Critical Value Test. It's your choice (or your teacher's), and we'll walk you through both.

These are the Step 4 action items that we'll cover:

1. Calculate the p-value and test statistic.
2. Complete the curve sketch.
3. Do the P-Value Test.
4. Do the Critical Value Test.

Calculate the P-Value and Test Statistic

To calculate the p-value and test statistic, we first need to find our observation. For questions about a population mean, it's usually straightforward, and we pull the sample mean straight from the question without doing any calculations.

For questions about population, however, there are a variety of ways that the information can be provided, which can be confusing. Sometimes, we're lucky and are given the proportion in the form of a decimal, percent, or proportion. Otherwise, we're given the part and the whole that, when put together, give us a proportion:

Identify the sample statistic.

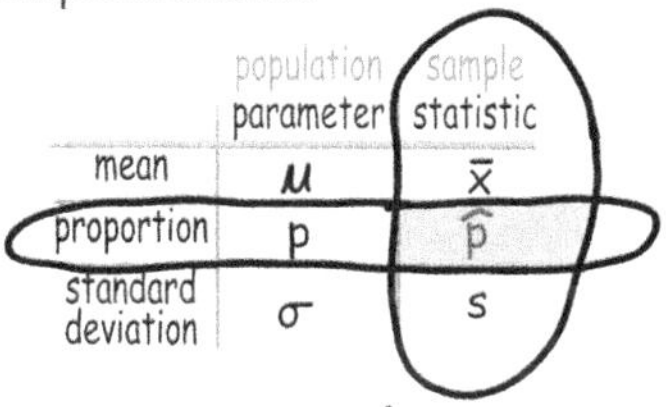

The observed sample statistic is $\hat{p}$, a sample proportion.

$$\hat{p} = \frac{\text{number of students w/30K or less debt}}{\text{total number of students in the sample}}$$

Let's take a look at our question to determine what our sample proportion is:

A newspaper article published that, in the United States, fifty percent of college students graduate with $30,000 or less in student loans. A college wants to know if their recent graduates differ from the national population; they claim that their graduates are different. They surveyed a simple random sample of 2,113 of their recent graduates. They found that 1,099 of them had a total of $30,000 or less in loan debt. Test the claim at a 0.05 significance level.

total number of students in the sample

number of students w/30K or less debt

$$\hat{p} = \frac{1099}{2113} \approx 0.52011$$

This is the observation, or the observed sample statistic in the observation space.

Now that we have the observation, we can calculate the p-value and the test statistic. The simplest way to do this is to use the

TI-84 test function 5:1-PropZTest. Here are the specific steps to do this:

• Choose Stat

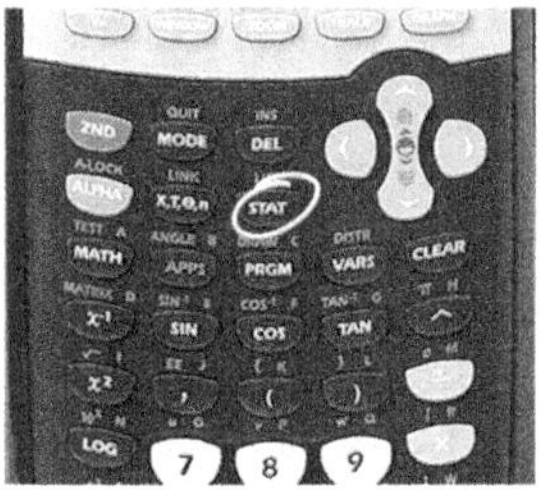

• Choose Test, then 5: 1-PropZTest

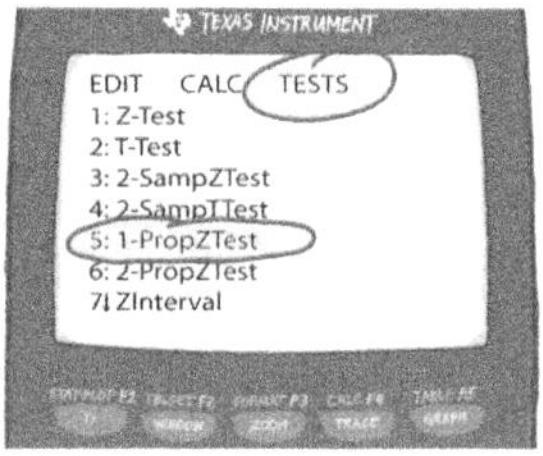

Here are the specific inputs you'll use:

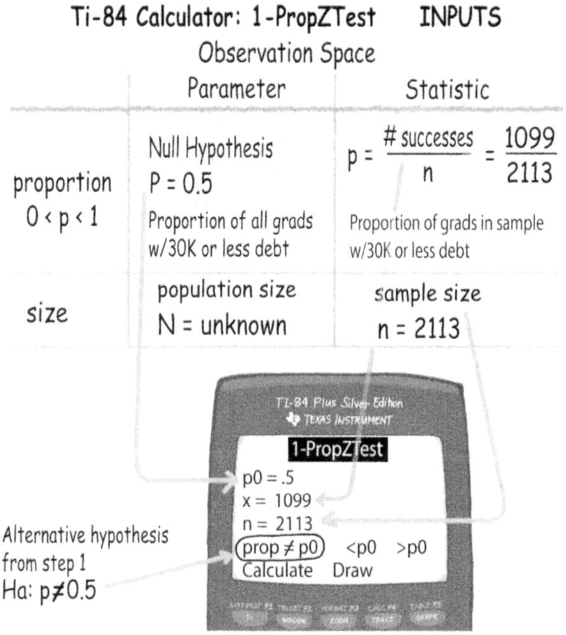

Confusing Notation: A point of confusion for many students is that the calculator uses slightly different notation than their teacher and book. Here is what each input refers to in our question:

- p0 refers to our null hypothesis, H_0: p = 0.5.
- x refers to our number of successes in the trial, 1099
- n refers to n, our sample size, 2113
- prop refers to our alternative hypothesis, H_a: p ≠0.5.

The next step is to interpret the outputs from the 1-PrepZTest:

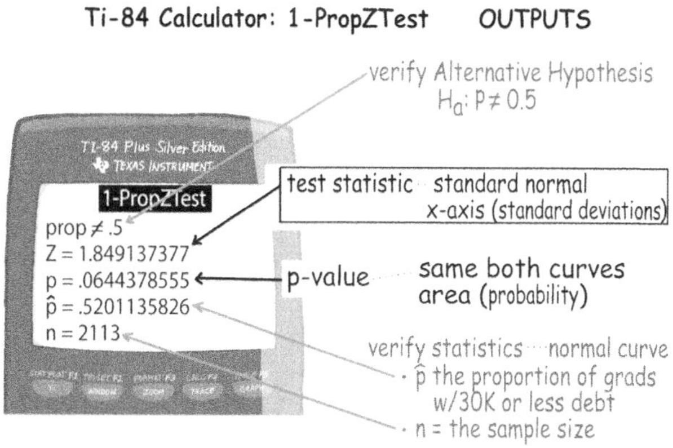

The p-value is p = 0.0644378555 and the test statistic is z = 1.849137377.

Complete the Curve Sketch

Next, we add the observation and the test statistic to their curves. Remember that the observation goes in the observation space (left graph), and the test statistic goes in the standardized space (right graph).

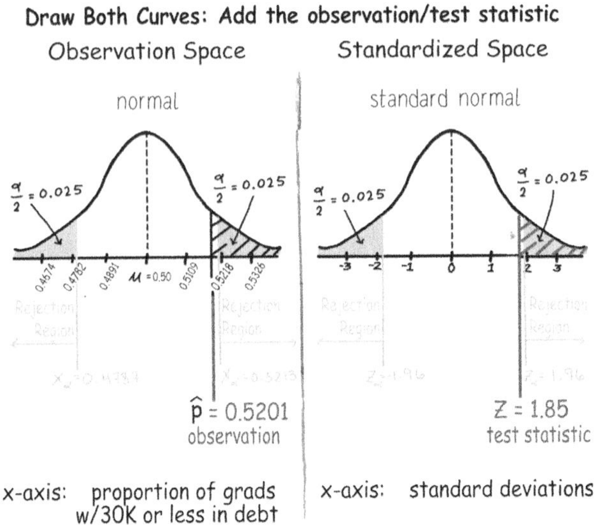

After adding the observation and test statistic to the appropriate curve, there is one additional step. Since this is a two-tailed test, we should reflect the observation and test statistic into the opposite tail to complete the curve sketch.

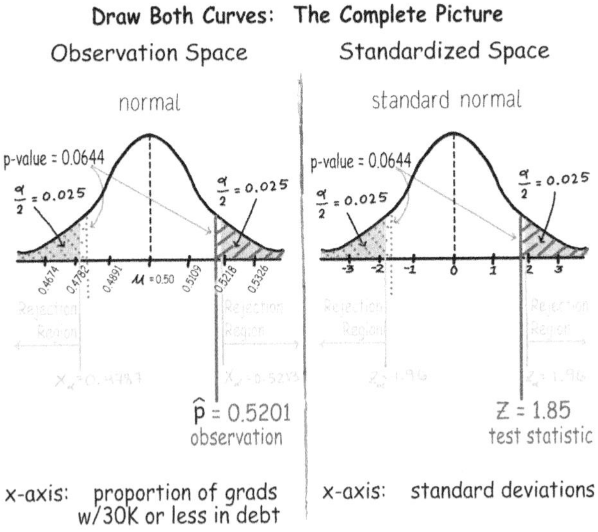

Common Question: How do I know where to draw the observation and test statistic on the sketched curve picture? This is where our original Step 3: Sketch of the Curves comes in handy. If we labeled our x-axis correctly, we can see where to place the observation and test statistic. See the graph below for placement.

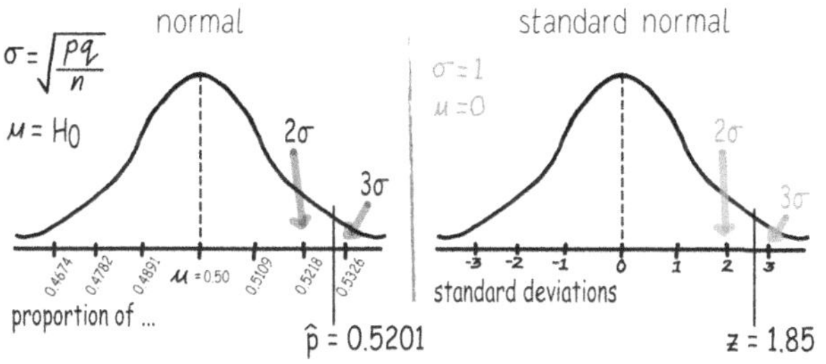

Do the P-Value Test

The P-Value Test compares two areas; the comparison of two areas tells us whether or not the test statistic is significant. If the area in the tail of the test statistic is smaller than the preselected significance level, the test statistic is significant. When we say it is significant, we mean that the test statistic is so unusual that we suspect it was not generated by a population with the mean used to create the curve in the observation space.

The P-Value Test states:

> P-value low, the null must go
> P-value $\le \alpha \Rightarrow$ Reject H_0
>
> P-value high, the null will fly
> P-value $> \alpha \Rightarrow$ Fail to Reject H_0

Remember, it is a two-tailed test. Therefore it is the area in both tails. It does not matter whether we are in the observation space or the standardized space. In both spaces, the p-value is the same size.

P-Value Method: Compare Areas

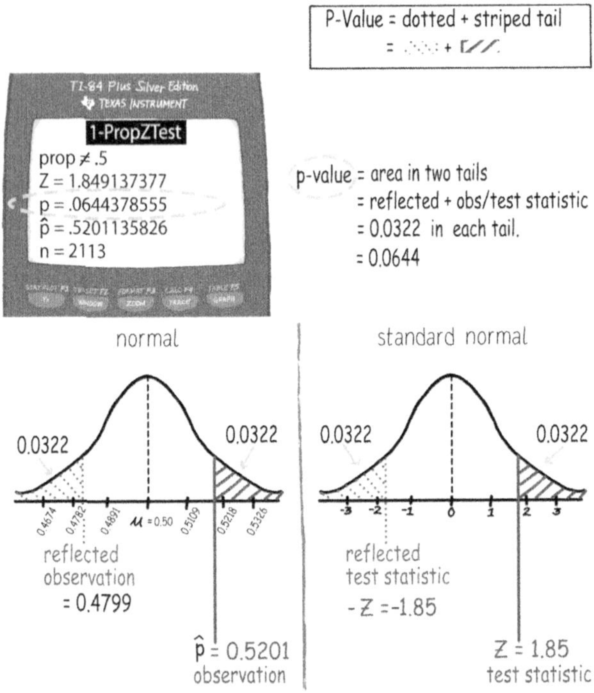

	Sketching Curves Checklist		
	CHECKLIST	OBSERVATION SPACE	STANDARDIZED SPACE
5.	significance level α (area) ☑	same area, both curves α = 0.05	Left Tailed / Right Tailed / (Two Tailed)
6.	critical value (x-axis) ☑	$X_{\alpha/2}$ left = 0.4786 right = 0.5213	$Z_{\alpha/2}$ $-Z_{\alpha/2} = -1.96$ $Z_{\alpha/2} = 1.96$
1.	observation/ test statistic (x-axis) ☑	observation $\hat{p} = \frac{\# \text{successes}}{n} = \frac{1099}{2113}$	test statistic $Z = 1.85$
2.	p-value (area) ☑	same area, both curves p-value $= 2 * 0.0322 = 0.0644$	Left Tailed / Right Tailed / (Two Tailed)

Compare the p-value area to the significance level area. Look at the complete picture of both curves. Note that the p-value area is larger than the solid-colored significance level area.

Sometimes, it's easier to see that in the curve sketch; other times, it's easiest to just compare the numeric values in the chart. If the p-value is less than or equal to **α**, we reject H_0. If the p-value is larger, we fail to reject H_0.

For this problem, we found that the p-value is 0.0644, and the significance level area is 0.05.

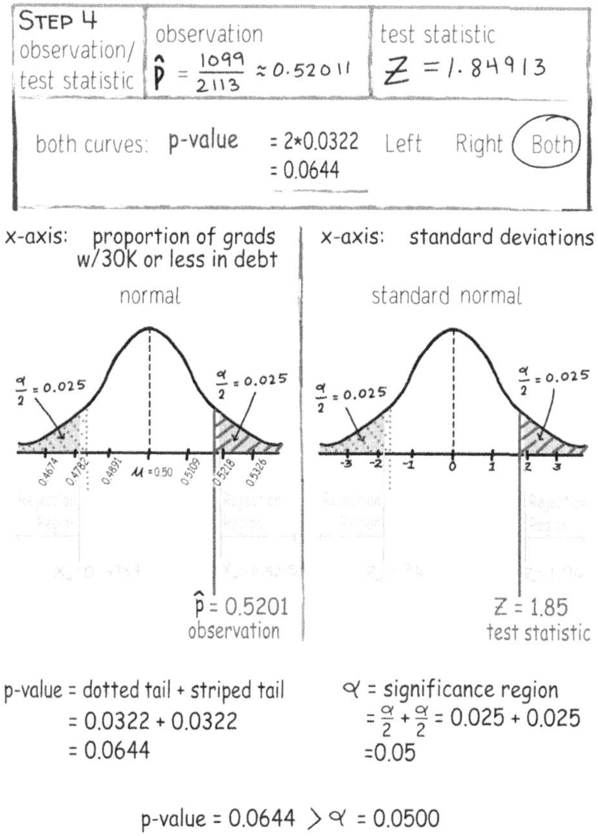

The p-value is larger, so we fail to reject the null hypothesis.

Step 4: Do the Test

TEST	P-VALUE METHOD
CALCULATE	P-Value = 2*0.0322 = 0.0644
TEST	Is P-Value ≤ α ? p-value = 0.0644 > α =0.05 (No) / Yes
DECISION	No → Fail to Reject H_0; Yes → Reject H_0

Do the Critical Value Test

Let's review the criteria for the Critical Value Test:

- When the test statistic falls within the rejection region interval, we reject the null hypothesis, H_0.
- When the test statistic does *not* fall within the rejection region, we fail to reject the null hypothesis, H_0.

We can determine this by looking at the x-axis of each of our graphs:

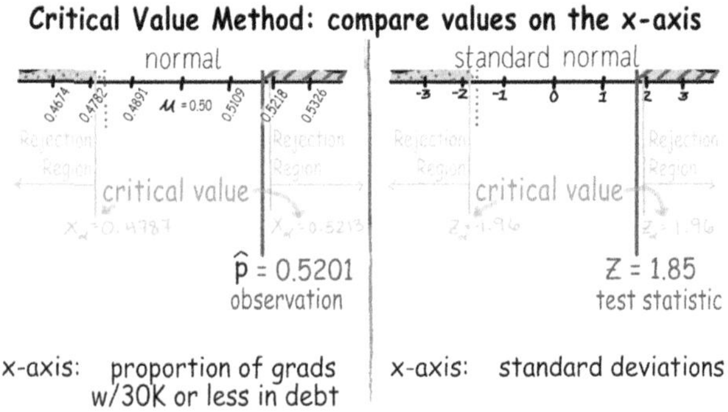

Let's look at the close-up of the axes. On the normal curve (on the left), the observation is 0.5201. It is a two-tailed test, so the rejection region is found in both tails. The rejection region intervals are: (-∞, 0.4787] and [0.5201, ∞). The observation is not contained in either rejection region interval, so we fail to reject the null hypothesis.

On the standard normal curve (on the right), the test statistic is z = 1.85, and we want to determine if it is inside either rejection region interval. There are two rejection region intervals, (-∞, -1.96] and [1.96, ∞). The test statistic is not found in either interval; the conclusion is that we fail to reject the null hypothesis.

We see that both curves, the normal and the standard normal, lead to the same conclusion.

Full Step 4 Solution

Check Your Work Tip: The P-Value Test and the Critical Value Test will always give you the same results. If your teacher requires you to do the Critical Value Test, but you're more comfortable with the P-Value Test, you can always use the P-Value Test to check your work for the Critical Value Test!

Sketching Curves Checklist

	CHECKLIST	OBSERVATION SPACE	STANDARDIZED SPACE
1.	distribution ☑	(normal circled) / Student t	(standard normal circled) / standardized Student t
2.	units (x-axis) ☑	proportion of grads w/30k or less debt	NO CHOICE standard deviations
3.	mean of curve ☑	μ = Null Hypothesis $p = 0.50$	NO CHOICE $\mu = 0$
4.	standard deviation of curve ☑	$\sqrt{\frac{pq}{n}} = \sqrt{\frac{(.5)(.5)}{2113}}$ ≈ 0.01088	circle one (1 circled) or $\approx$ 1
5.	significance level α (area) ☑	same area, both curves $\alpha = 0.05$	Left Tailed / Right Tailed / (Two Tailed circled)
6.	critical value (x-axis) ☑	$X_{\alpha/2}$ left = 0.4786 right = 0.5213	$Z_{\alpha/2}$ $-Z_{\alpha/2} = -1.96$ $Z_{\alpha/2} = 1.96$
1.	observation/ test statistic (x-axis) ☑	observation $\hat{p} = \frac{\#\text{ successes}}{n} = \frac{1099}{2113}$	test statistic $Z = 1.85$
2.	p-value (area) ☑	same area, both curves p-value $= 2 * 0.0322 = 0.0644$	Left Tailed / Right Tailed / (Two Tailed circled)

Draw Both Curves: The Complete Picture

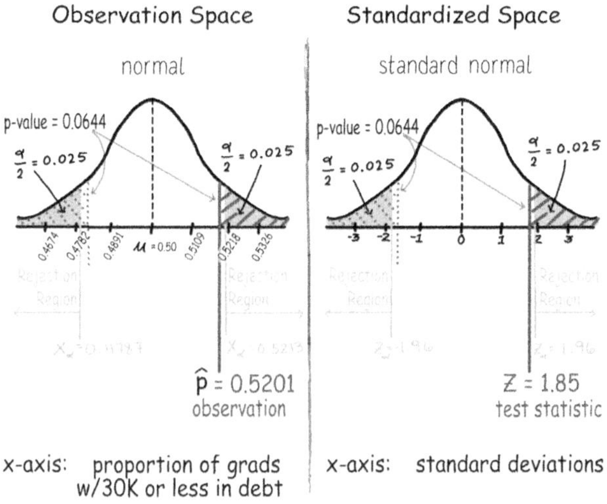

Filling in our P-Value Test and Critical Value Test results from earlier; we see that both tests reach the same conclusion.

STEP 4: DO THE TEST

TEST	P-VALUE METHOD	CRITICAL VALUE METHOD
CALCULATE	P-Value = 2*0.0322 = 0.0644	Test Statistic =1.85
TEST	Is P-Value ≤ α ? p-value = 0.0644 > α =0.05 (No) Yes	Is Test Statistic in the Rejection Region? (No) Yes
DECISION	(Fail to Reject H_0) Reject H_0	(Fail to Reject H_0) Reject H_0

Check Your Work: Both the observation and test statistic represent the same thing; the only difference is that one is in the observation space and the other is in the standardized space. Because they represent the same thing, they are drawn in the same location relative to the appropriate curve; both will be placed between the second and third standard deviations above the mean. Why? The two graphs are simply different ways of representing the same thing:

- On the normal curve, P-hat = 0.5201 is 1.85 standard deviations above the mean, $\mu = 0.05$.
- On the standard normal curve, z = 1.85 is 1.85 standard deviations above the mean, $\mu = 0$.

Visually, you should be able to inspect the two graphs and find the observation and test statistic in roughly the same place. If they aren't, go back and check your work (and be thankful that you drew the graphs because otherwise, you might not have caught your mistake.)

Common Question: What if I am given the test statistic and asked to find the p-value? In straightforward 1-Sample Population Proportion questions, the 1-PropZTest() function will suffice. You might, however, get a question where you must use the normalcdf() function to find the p-value. We'll cover that here as well.

If your teacher gives you the test statistic, Z, and asks you to find the corresponding p-value, here is how you'll do it. On your calculator, you'll choose 2ND → VARS→ 2:normalcdf().

1) Choose 2nd, then Distr 2) Choose 2: normalcdf

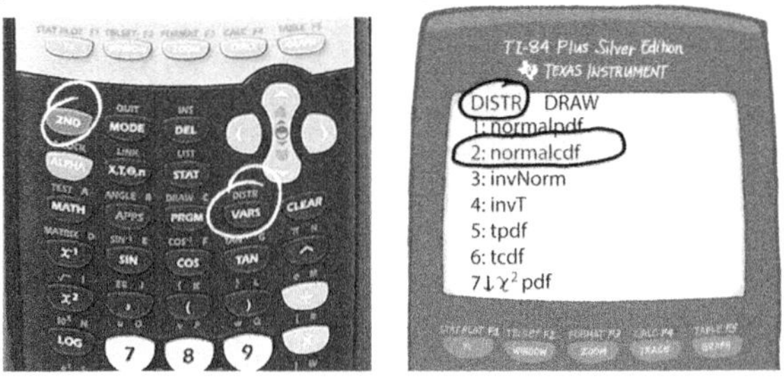

The normalcdf() function requires the following inputs:

- Lower: This is the lower bound of your shaded area.
- Upper: This is the upper bound of your shaded area.
- μ & σ: Also pull these from your graph.

When done, hit PASTE, then ENTER.

Since this is a two-tailed problem, we'll need to find the area in the left tail and the area in the right tail. (Or, because the tails are symmetric, we can find the area in one tail, then multiply by 2.)

Here's how we take the information from the graph to create the calculator inputs:

3) Add inputs, then hit enter

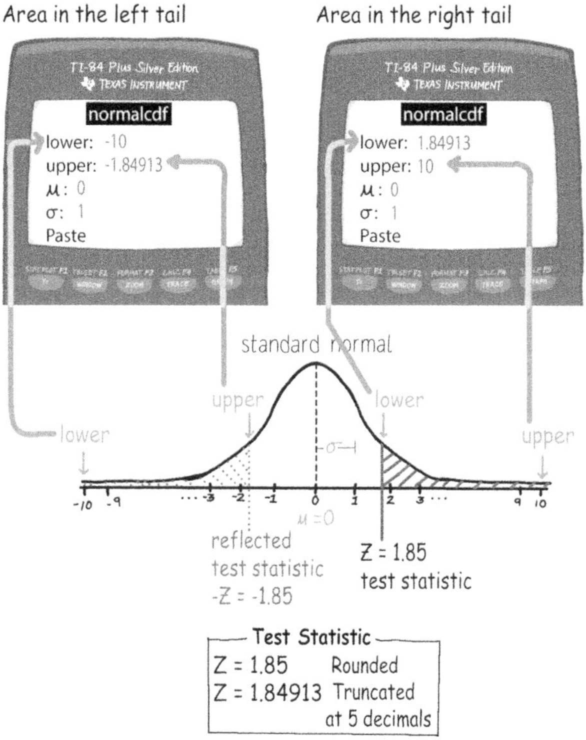

Rounding Tip: When you first get calculator outputs that require rounding, write down both the rounded *and* the full non-rounded number. When we label our graphs, we round to two decimal places just because it's neater. However, when we input into the normalcdf() or other calculator functions, it's best to use the full non-rounded number or the number rounded to five decimal places.

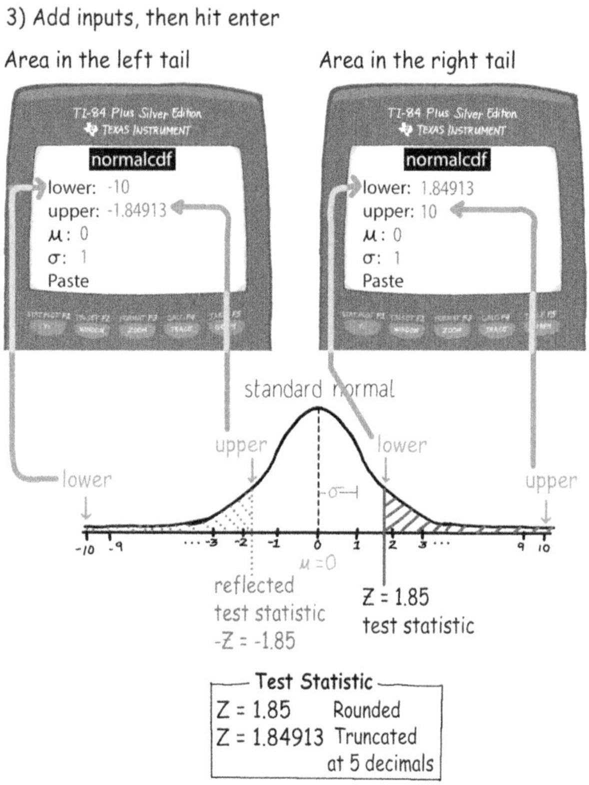

Once we have our inputs, we hit PASTE, then ENTER, and look at the outputs:

4) Interpret Outputs

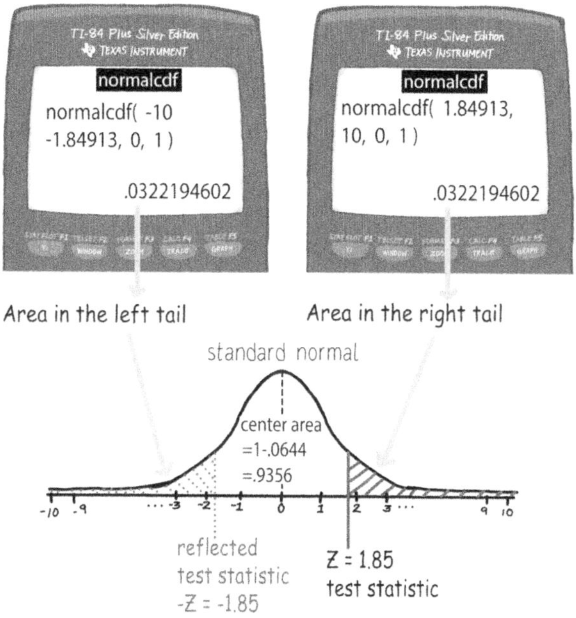

P-Value = dotted area left tail + striped area right tail
= 2 * 0.03221
= 0.06442
round the final answer to four digits to obtain
P-Value = 0.0644

Our p-value is 0.0644, matching exactly what we got when we used the 1-PropZTest.

Here's a quick recap:

Sketching Curves Checklist			
	CHECKLIST	OBSERVATION SPACE	STANDARDIZED SPACE
1.	observation/ test statistic ☑ (x-axis)	observation $\hat{p} = \frac{\#\text{ successes}}{n} = \frac{1099}{2113}$	test statistic $Z = 1.85$
2.	p-value ☑ (area)	same area, both curves p-value =2*0.0322 =0.0644	Left Tailed Right Tailed (Two Tailed)

A+ Student Questions:

- For a two-tailed test, how do I find both tails when I only have one test statistic?
- Given the observation, how do I find the test statistic using a formula (not the calculator)?
- Given the test statistic, how do I find the observation using algebra (not technology)?

You can find answers to the above questions, along with answers to other *A+ Student Questions*, at allisonlovesmath.com/hypothesis-testing.

18. STEP 5 STATE THE CONCLUSION

According to both test methods, we should fail to reject the null hypothesis, H_0. Now, our last step is to state our conclusion.

Many students lose points on their hypothesis testing exam by botching up the final conclusion. Yes, Statistics is a math class, but it requires strong reading and writing skills. If writing mathy conclusion sentences scares you, the only thing I can say is that you have to toughen up and work through the fear. If Statistics throws some tricky wording into the final step of the problem, don't be the student who gives up. Acknowledge that it's hard and that, perhaps, such writing is a new skill for you, then commit to figuring it out!

Here's how Step 5: State the Conclusion breaks down:

1. Follow the Flow Chart.
2. Write the Basic Conclusion.
3. Write the Meaningful Conclusion.

Try to do Step 5 yourself first, then read through the final answer below.

Follow the Flow Chart

To help you out with the wording, we created a handy flowchart for you. Circle your Step 1, Circle your Step 4, and the initial wording for your conclusion is right there for you!

STEP 5: STATE YOUR CONCLUSION

Step 1: Claim	Step 4: Result	Conclusion
Claim is H_A	→ Reject H_0 →	There is sufficient evidence to support the claim that....
	→ Fail to Reject H_0 →	There is not sufficient evidence to support the claim that....
Claim is H_0	→ Reject H_0 →	There is sufficient evidence to warrant the rejection of the claim that....
	→ Fail to Reject H_0 →	There is not sufficient evidence to warrant the rejection of the claim that....

Confused about how we filled out the chart above? First, we referred back to Step 1:

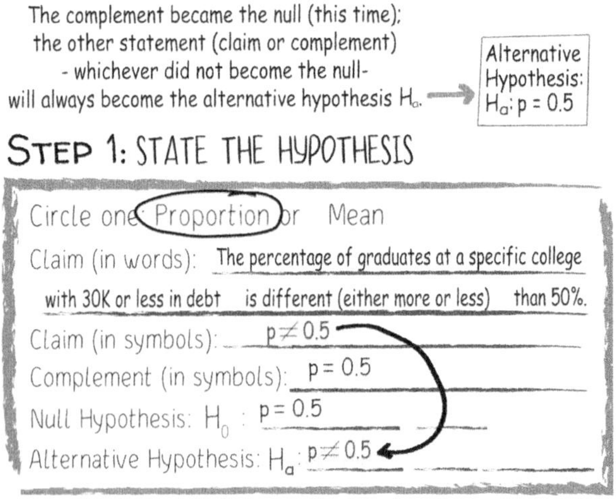

As you can see from our original Step 1, our claim is our alternative hypothesis, H_a.

Second, we refer back to Step 4,

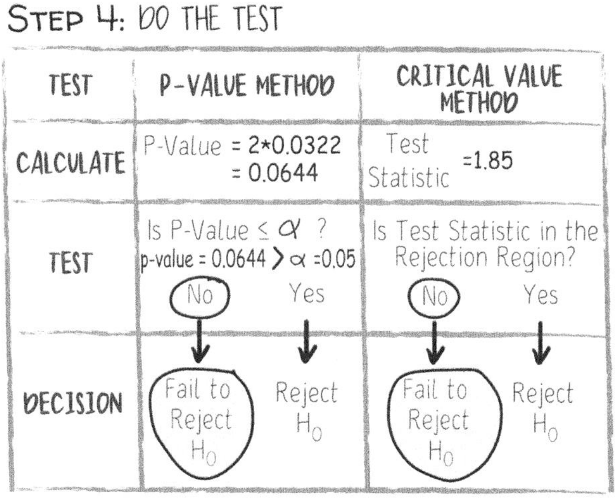

Regardless of which method you chose, both the P-Value Test and the Critical Value Test resulted in Failing to Reject the Null Hypothesis, H_0.

Write the Basic Conclusion

To write a basic conclusion, take the wording from the flow chart and combine it with the original wording in the claim:

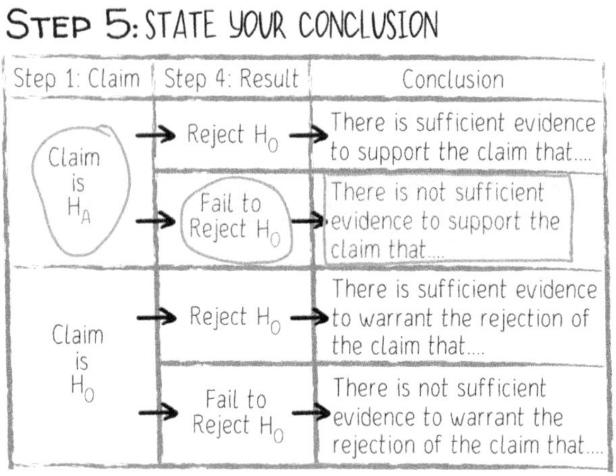

"There is not sufficient evidence to support the claim that…"

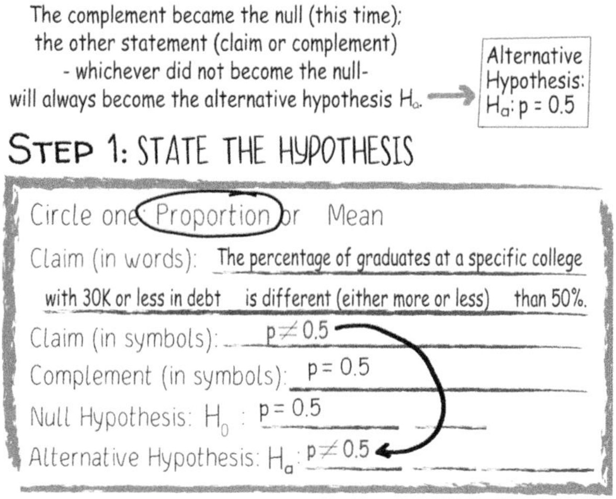

"The percentage of graduates at a specific college with \$30,000 or less debt is different than 50%."

Putting these two together, we get:

"There is not sufficient evidence to support the claim the percentage of graduates at a specific college with \$30,000 or less debt is different than 50%."

Write the Meaningful Conclusion

Of course, the problem with the basic conclusion we just covered is that it often doesn't mean anything to students. You're just following a flow chart and not thinking about what it means. That is why many teachers add a question asking you to explain what it means in your own words.

To do this, let's refer back to our original questions.

> A newspaper article published that, in the United States, fifty percent of college students graduate with $30,000 or less in student loans. A college wants to know if their recent graduates differ from the national population; they claim that their graduates are different. They surveyed a simple random sample of 2,113 of their recent graduates. They found that 1,099 of them had a total of $30,000 or less in loan debt. Test the claim at a 0.05 significance level.

What is the question about? A college wants to know if their recent graduates are different. Do 50% of their recent graduates also graduate with $30,000 or less of student debt or not?

How do we interpret this basic conclusion?

"There is not sufficient evidence to support the claim the percentage of graduates at a specific college with $30,000 or less debt is different than 50%."

The basic conclusion is intended to help you get your thinking straight. Your job is not done until you add another sentence restating the conclusion in plain English, in your own words. The full conclusion should contain, at a minimum, a second sentence that is meaningful in the context of the initial problem statement. What was the question asking about? *A college wants to claim their graduates leave with loan debt that is different from the national average.* Use that in the conclusion.

Many different correct conclusions can be written. Notice that the two different conclusions presented here assume a slightly different context. Both are correct.

Conclusion Option 1: There is not sufficient evidence to support the claim that the percentage of graduates at the specified college who graduate with $30,000 or less in debt is different than 50%. The debt burden of recent graduates from this college does not appear to be different from the debt burden of students nationwide.

Conclusion Option 2:

The percentage of graduates at the college with $30,000 or less debt does not appear to be different from 50%."

The college can assume that its recent graduate student debt amounts are similar to others in the United States and can proceed accordingly. If their goal is to have student debt less than the typical recent grad, they'll take measures to decrease student debt. If they simply want to make sure that they aren't above average, they may simply say, "Yay, we don't have to worry about this problem now." Either way, the college has the information needed to inform their actions.

Warning: There are many possible correct conclusions, but there are even more incorrect options. When writing the conclusion in your own words, make sure that what you write agrees with the scripted standard conclusion and the original problem statement.

Wow, that was a deep dive into how to do a hypothesis test for 1-Sample Population Proportions. We dug deep into two-tailed tests also. In the next section, on one-sample means, we'll take a deeper look at a 1-sample test.

Hopefully, that breakdown was helpful for you, answered some of your questions, and got you through the first part of your hypothesis testing coursework.

PART IV

HYPOTHESIS TESTING FOR A 1-SAMPLE POPULATION MEAN

In this chapter, we walk you step by step through how to solve a 1-sample hypothesis testing question about a population mean.

Step 1: State the Hypotheses. Here, we show you what words in the prompt indicate that a question is asking about a population mean (and not a population proportion) and how to correctly state your null and alternative hypotheses.

Step 2: Choose a Distribution. We'll show you how to know which test to use and do a quick comparison between the tests for population proportions versus population means.

Step 3: Sketch the Curves is (still!) the most important of the five steps. Do not give in to the temptation to skip it! In this step, you'll learn how to graph the Student t curve and the standardized Student t curve. We'll also calculate and label the area for α for the P-Value Test and the critical value and rejection region for the Critical Value Test.

Step 4: Do the Test. We teach you the individual hypothesis tests: the Critical Value Test and the P-Value Test. While there is much overlap between the two, it's helpful to see each test broken down individually.

Step 5: State the Conclusion. We walk you through the flow chart to figure out what wording to use in your final conclusion and share some common mistakes for writing a good conclusion.

19. THE 1-SAMPLE POPULATION MEAN QUESTION

Here's the 1-sample population proportion question we'll be tackling:

> A college wishes to monitor the financial health of its graduates. According to the 8% rule, for graduates earning $30,000 a year, a monthly payment of $200 or more indicates a financial health problem. The college assesses loan payments for a simple random sample of its graduates earning $30,000 a year. The results of the study are $n = 40$, $\bar{x} = \$185$, and $s = \$50$.
> Use a 0.05 significance level to test the claim that this sample is from a population with good financial health.

As with the population proportions question in the last section, a ton of information is packed into these six lines. On top of that, students typically spend less time on the Student t distribution than the normal distribution, so there are several levels of anxiety-inducing wording packed into this question.

If this is stressing you out, take a deep breath, and remember that the point of all of this is to learn how to make sense of complicated, intimidating things. If it weren't hard, you wouldn't need a class to learn it.

So, how do we tackle challenges like this?

1. **Embrace active learning.** Try to solve the problem yourself first, then review the answers to learn from your mistakes. Learning from mistakes is much harder but more effective than simply reading through the answers without doing the work yourself.
2. **Take it one step at a time.** If you can't do the entire problem first, just try Step 1, and check your work. After you fully understand Step 1, try Step 2, and check your work.

These two steps are the keys to learning for anything difficult, including hypothesis questions.

Try to work through the question on your own. Then, check your work on the next page.

Solution

After you try the question, check your answer. Below is the quick answer.

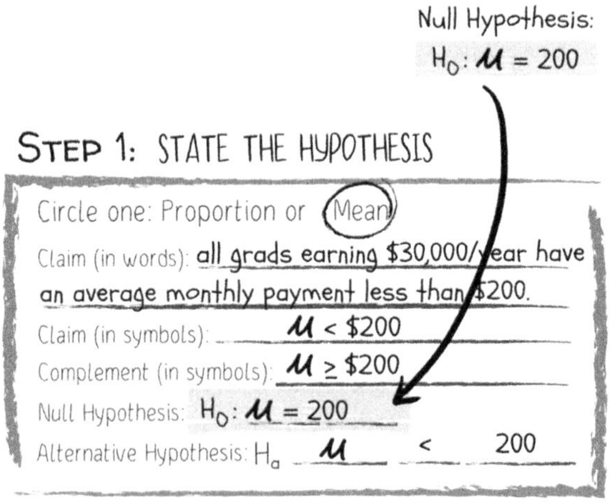

STEP 2: Choose A Distribution

	PROPORTION	MEAN
REQUIREMENTS	1. SRS 2. Binomial a. Fixed # trials b. Trials independent c. Two outcomes d. Fixed Probablity 3. 5 successes /5 failures	1. SRS ✓ 2. Population is normally dist or n>30 ✓ 3. σ known or σ unknown ✓
	NORMAL	NORMAL / STUDENT - T

STEP 3	OBSERVATION SPACE	STANDARDIZED SPACE		
distribution	normal / Student t	standard normal / standardized Student t		
x-axis units/ label	dollars	NO CHOICE standard deviations		
mean of curve	μ = Null Hypothesis $\mu = 200$	NO CHOICE $\mu = 0$		
standard deviation of curve	$\frac{s}{\sqrt{n}} = \frac{50}{\sqrt{40}} \approx 7.9$	circle one 1 or ≈ 1		
both curves: significance level α = 0.05		Left Tailed	Right Tailed	Two Tailed
critical value	$X_\alpha = 186.68$	$t_\alpha = -1.68$		
STEP 4 observation/ test statistic	observation $\bar{x} = 185.00$	test statistic $t = -1.90$		
both curves: p-value = 0.0326		Left Tailed	Right Tailed	Two Tailed

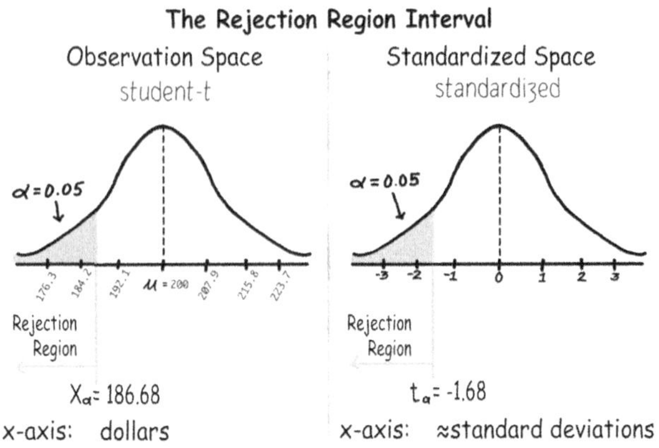

For STEP 3: Sketch the Curves, fill in the following checklist as you go to track what you need to label on the curves.

STEP 3	OBSERVATION SPACE	STANDARDIZED SPACE
distribution	normal / Student t	standard normal / standardized Student t
x-axis units/ label	dollars	NO CHOICE standard deviations
mean of curve	μ = Null Hypothesis $\mu = 200$	NO CHOICE $\mu = 0$
standard deviation of curve	$\frac{s}{\sqrt{n}} = \frac{50}{\sqrt{40}} \approx 7.9$	circle one 1 or ≈ 1
both curves: significance level α = 0.05		Left Tailed Right Tailed Two Tailed
critical value	$X_\alpha = 186.68$	$t_\alpha = -1.68$
STEP 4 observation/ test statistic	observation $\bar{X} = 185.00$	test statistic $t = -1.90$
both curves: p-value = 0.0326		Left Tailed Right Tailed Two Tailed

Here are the fully filled out graphs for Step 3: Sketch the Curve:

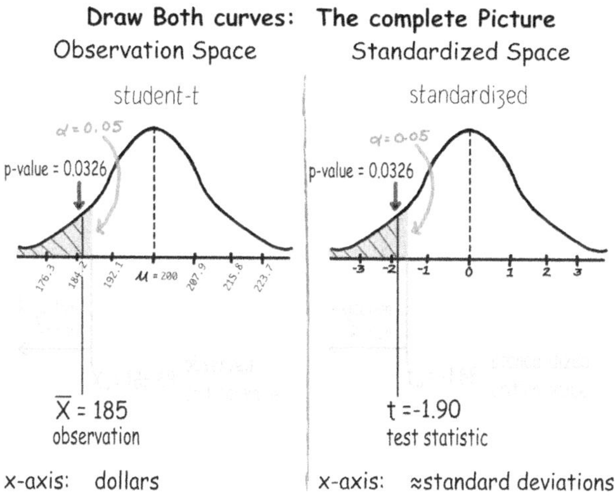

Here are the final calculations, graphs, and decisions for Step 4: Do the Test:

STEP 3	OBSERVATION SPACE	STANDARDIZED SPACE
distribution	normal / **Student t**	standard normal / **standardized Student t**
x-axis units/ label	dollars	NO CHOICE standard deviations
mean of curve	μ = Null Hypothesis $\mu = 200$	NO CHOICE $\mu = 0$
standard deviation of curve	$\frac{s}{\sqrt{n}} = \frac{50}{\sqrt{40}} \approx 7.9$	circle one 1 or **≈ 1**
both curves: significance level α = 0.05		**Left Tailed** Right Tailed Two Tailed
critical value	$X_\alpha = 186.68$	$t_\alpha = -1.68$
STEP 4 observation/ test statistic	observation $\bar{x} = 185.00$	test statistic $t = -1.90$
both curves: p-value = 0.0326		**Left Tailed** Right Tailed Two Tailed

TEST	P-VALUE METHOD	CRITICAL VALUE METHOD
CALCULATE	P-Value = 0.0326	Test Statistic = -1.90
TEST	Is P-Value ≤ α ? 0.0326 ≤ 0.05 No / **Yes**	Is Test Statistic in the Rejection Region? No / **Yes**
DECISION	Fail to Reject H_0 / **Reject H_0**	Fail to Reject H_0 / **Reject H_0**

STEP 5: STATE YOUR CONCLUSION

Step 1: Claim	Step 4: Result	Conclusion
Claim is H_A	→ Reject H_0 →	There is sufficient evidence to support the claim that....
	→ Fail to Reject H_0 →	There is not sufficient evidence to support the claim that....
Claim is H_0	→ Reject H_0 →	There is sufficient evidence to warrant the rejection of the claim that....
	→ Fail to Reject H_0 →	There is not sufficient evidence to warrant the rejection of the claim that....

Conclusion: There is sufficient evidence to support the claim that recent graduates come from a population with a mean monthly payment of less than \$200, which is considered too low relative to their \$30,000 a year salary.

If you got it wrong, no worries! Read through the following sections and make notes. What did you miss? What did you do correctly? What were the sneaky traps you should look out for on your test? Tracking all these things will help make studying for your hypothesis testing exam easier.

If you got the answer correct, good job! If skimming through the pages can help you to understand why you got the correct answer, build your confidence in what you're doing, and fill in any gaps you may have in your understanding of the question. Go for it!

Let's go through this in more detail, step-by-step.

20. STEP 1: STATE THE HYPOTHESES

The Step 1 building blocks:

1. State the parameter.
2. State the claim and complement.
3. Write the hypotheses.

Step 1 is the complete story of how to identify the precise words that mold the claim and the complement and then how those two statements are written symbolically and transformed into the null and alternative hypotheses.

State the Parameter

The first step is to read the problem and identify whether it is about a proportion or a mean.

A college wishes to monitor the financial health of its graduates. According to the 8% rule, for graduates earning $30,000 a year, a monthly payment of $200 or more indicates a financial health problem. The college assesses loan payments for a simple random sample of its graduates earning $30,000 a year. The results of the study are $n = 40$, $\bar{x} = \$185$, and $s = \$50$.
Use a 0.05 significance level to test the claim that this sample is from a population with good financial health.

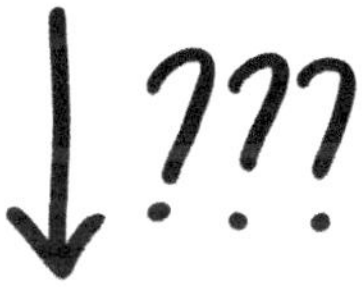

STEP 1: STATE THE HYPOTHESIS

Circle one: Proportion or Mean

Claim (in words): ______________________________

____

Claim (in symbols): ____________________________

Complement (in symbols): _______________________

Null Hypothesis: H_0 : ______ ____

Alternative Hypothesis: H_a ______ ____ ______

Like percentage or ratio, keywords often indicate a proportion, while sample data containing a mean will usually indicate a mean. Highlight the population and the keywords. Circle the keywords in the problem statement.

What do we do if we have a percent and a sample mean? Read the claim and determine if the question is about the percent or the mean. The claim we have underlined is asking about "a population with good financial health." What does it mean to have good financial health?

A college wishes to monitor the financial health of its graduates. According to the 8% rule, for graduates earning \$30,000 a year, a monthly payment of \$200 or more indicates a financial health problem. The college assesses loan payments for a simple random sample of its graduates earning \$30,000 a year. The results of the study are $n = 40$, $\bar{x} = \$185$, and $s = \$50$.
Use a 0.05 significance level to test the claim that this sample is from a population with good financial health.

In the problem statement, a financial health problem is defined as a monthly payment of \$200 or more. Good financial health is the complement of bad financial health. The student must find the complement; it is not explicitly given. Good financial health is a payment that is less than \$200.

Let's revisit our initial question; is this a problem about a proportion or a mean? The question is fundamentally about the average monthly loan payment. This is a problem about a mean.

Common Mistake: Circling words that technically are proportions or means, but aren't related to the claim. This question is particularly sneaky because we added "the 8% rule," which is not related to the claim. It's just the title of the guidelines the college is following.

Common Mistake: Thinking the claim is about the sample mean. Whether it's a proportion or mean, the claim for a hypothesis test is about a population. Keywords like "results of the study" as well as the symbol used tell you that even though \$185 is a mean, it is from a sample and therefore not part of the claim.

A college wishes to monitor the financial health of its graduates. According to the 8% rule, for graduates earning \$30,000 a year, a monthly payment of \$200 or more indicates a financial health problem. The college assesses loan payments for a simple random sample of its graduates earning \$30,000 a year. The results of the study are $n = 40$, $\bar{x} = \$185$ and $s = \$50$.
Use a 0.05 significance level to test the claim that this sample is from a population with good financial health.

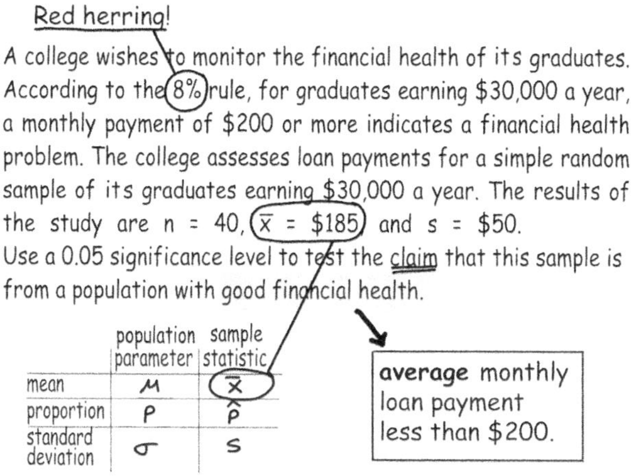

The claim in the problem statement is about an average. This is a problem about a mean.

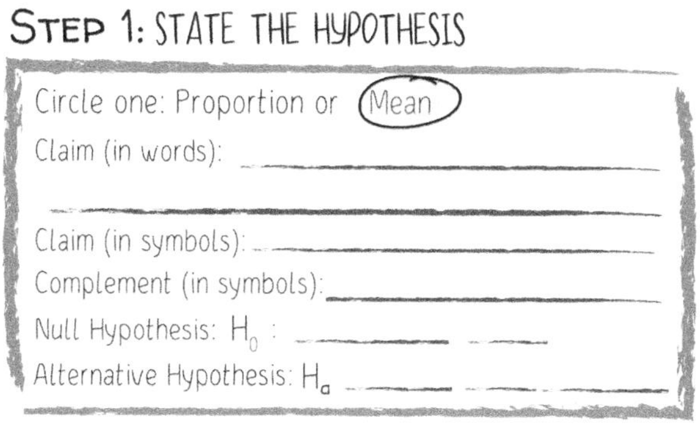

State the Claim and Complement

A claim is a hypothesis about a population. To write a claim, it is essential to identify two things—the population and then the specific claim about that population.

A college wishes to monitor the financial health of its graduates. According to the 8% rule, for graduates earning \$30,000 a year, a monthly payment of \$200 or more indicates a financial health problem. The college assesses loan payments for a simple random sample of its graduates earning \$30,000 a year. The results of the study are $n = 40$, $\bar{x} = \$185$, and $s = \$50$.
Use a 0.05 significance level to test the claim that this sample is from a population with good financial health.

Population	Sample
all graduates earning \$30K a year	graduates from a specific college earning \$30K a year

Now, let's identify the claim being made about this population.

A college wishes to monitor the financial health of its graduates. According to the 8% rule, for graduates earning \$30,000 a year, a monthly payment of \$200 or more indicates a financial health problem. The college assesses loan payments for a simple random sample of its graduates earning \$30,000 a year. The results of the study are $n = 40$, $\bar{x} = \$185$, and $s = \$50$.
Use a 0.05 significance level to test the claim that this sample is from a population with good financial health.

> The claim in the problem statement discusses a sample.
> Rewrite this to be a claim about a **population**.

Population	Rough Claim
all graduates earning \$30K a year	a population with good financial health

What is meant by good financial health?

A college wishes to monitor the financial health of its graduates. According to the 8% rule, for graduates earning $30,000 a year, a monthly payment of $200 or more indicates a financial health problem. The college assesses loan payments for a simple random sample of its graduates earning $30,000 a year. The results of the study are $n = 40$, $\bar{x} = \$185$, and $s = \$50$.
Use a 0.05 significance level to test the claim that this sample is from a population with good financial health.

Good financial health is the complement of a financial health problem.

financial health problem → a monthly payment of $200 or more indicates a financial health problem

good financial health → a monthly payment of less than $200 indicates good financial health

Let's put this all together and write the claim:

- Identify the claim in the context of the problem statement. The college claims that this sample is from a population with good financial health.
- Rewrite the claim to be about the population: =THE POP + THE CLAIM
- = (all graduates earning $30,000 a year) + (have good financial health)
- Simplified: All grads earning $30,000 a year have an average monthly payment of less than $200.

STEP 1: STATE THE HYPOTHESIS

Circle one: Proportion or (Mean)

Claim (in words): all grads earning $30K a year have an average monthly payment less than $200.

Claim (in symbols): ______

Complement (in symbols): ______

Null Hypothesis: H_0 : ______

Alternative Hypothesis: H_a ______

To write the claim symbolically, choose the population symbol from the chart.

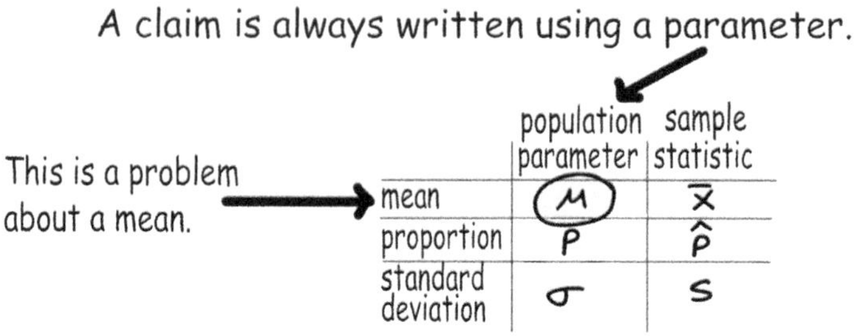

Finally, we can write the claim symbolically.

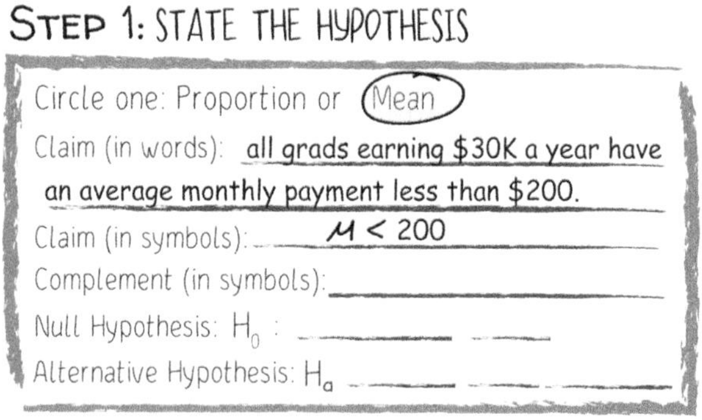

Common Mistake: Writing the claim with symbols for the statistic, rather than the population. Why is that wrong? Because our claim is always about the population. If we have a sample, we know what the mean of the sample is. We don't need to make claims about it or do hypothesis tests about it. It's when we have a population, a group that is too large to gather data from every person or thing in the set, that we need to make and test claims. Therefore, our claim is always about the population, and the mean of a population is μ.

Next, write out the complement of the claim. Remember, the

complement is the opposite of the claim. It's the statement that must be true if the claim is false.

Let's fill this all in on the worksheet.

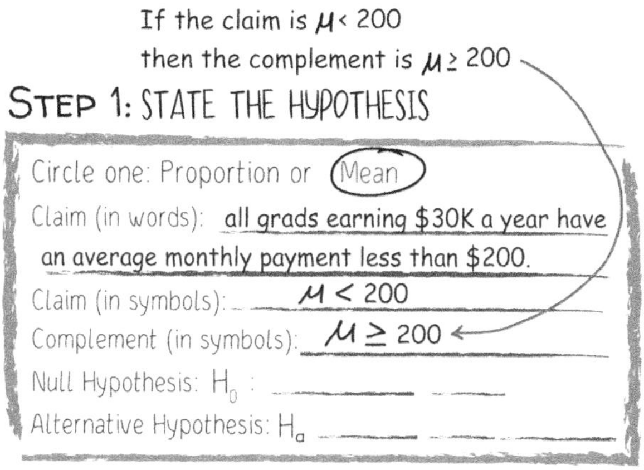

Common Mistake: Blindly turning the complement into the alternative hypothesis, H_a. Why is this a mistake? Because in some questions (like this one), the complement contains an equality statement; whichever statement contains the equality is the statement that will become the null hypothesis! The equality sign always tells us which statement will become the null hypothesis.

Write the Hypotheses

Next, let's write the null hypothesis! The complement is the equality statement; it must be rewritten as a strict equality to form the null hypothesis.

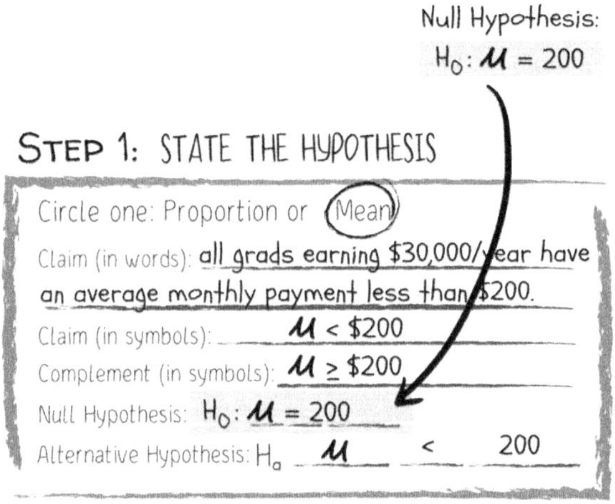

The complement became the null hypothesis; the other statement will become the alternative hypothesis.

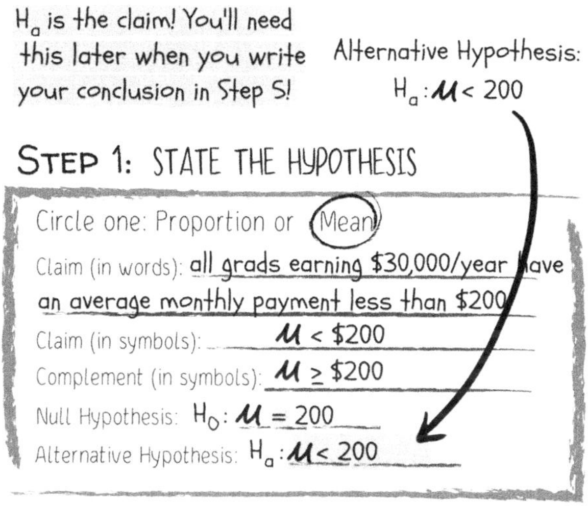

Woohoo! We're done with the setup and ready for Step 2!

21. STEP 2 CHOOSE A DISTRIBUTION

These are the Step 2 building blocks:

1. Restate the Parameter.
2. Check the Requirements.
3. Choose a Distribution.

Since we are only looking at two types of hypothesis tests (i.e., for population proportions and population means), there are only two options—the normal distribution or the Student t distribution.

Here is Step 2 already filled out, so you can check your work.

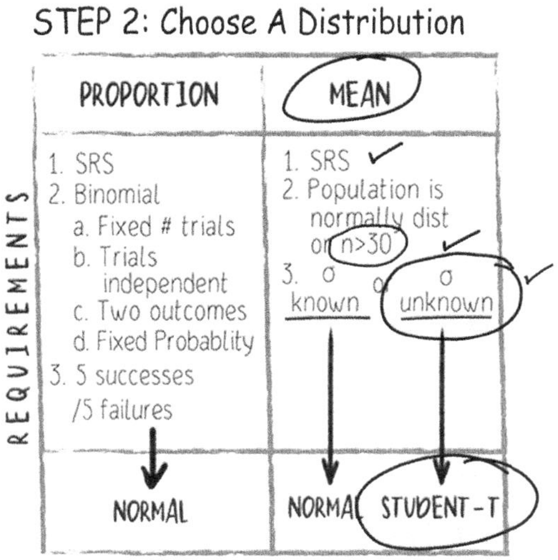

Restate the Parameter

The first thing Step 2 asks is, "Is our test about a population proportion or a population mean?" That's important because:

- If the test is about a population proportion, then we use the normal distribution.
- If the test is about a population mean, we usually use the normal or Student t distribution.

If you've done Step 1 correctly, you know this is a problem about a mean.

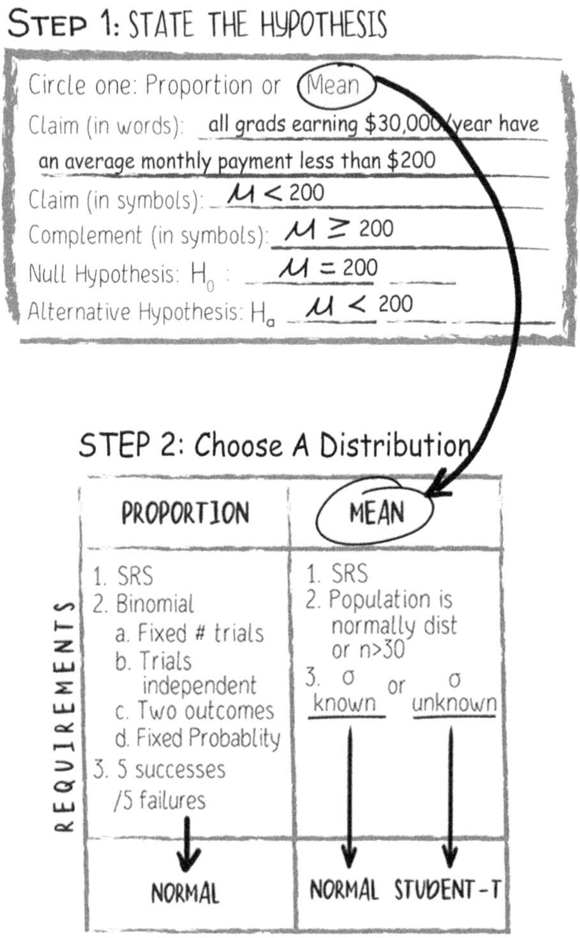

Done! It's nice to have an easy part once in a while.

Check the Requirements

Checking requirements is difficult, and introductory textbooks are notoriously bad about including all the information to show that requirements are met. Sometimes all the work is shown; other times, the textbook runs out of space, so this step is cut. But the student is expected to learn the step even when the textbook doesn't show the solution. We, as teachers, tell our students to always prove that the requirements are met, then a homework question might leave necessary information out. It's a pain.

That being said, requirements are important. In real-world applications, requirements must always be satisfied.

The first requirement for a hypothesis test about a population mean states that we need a simple random sample (SRS). Go back to the problem statement to identify keywords that indicate a simple random sample.

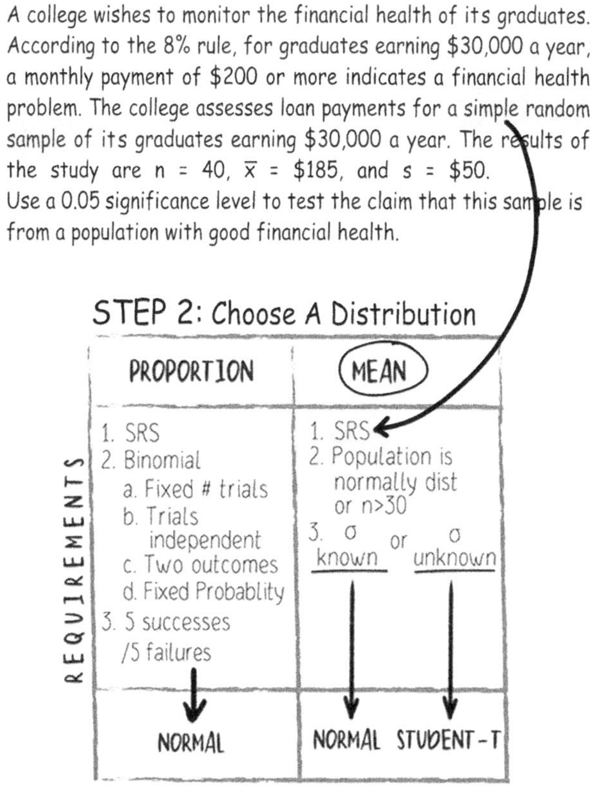

The second requirement is having a population that is normally distributed or large enough (n>30). Again, check the problem statement to determine if either requirement is satisfied.

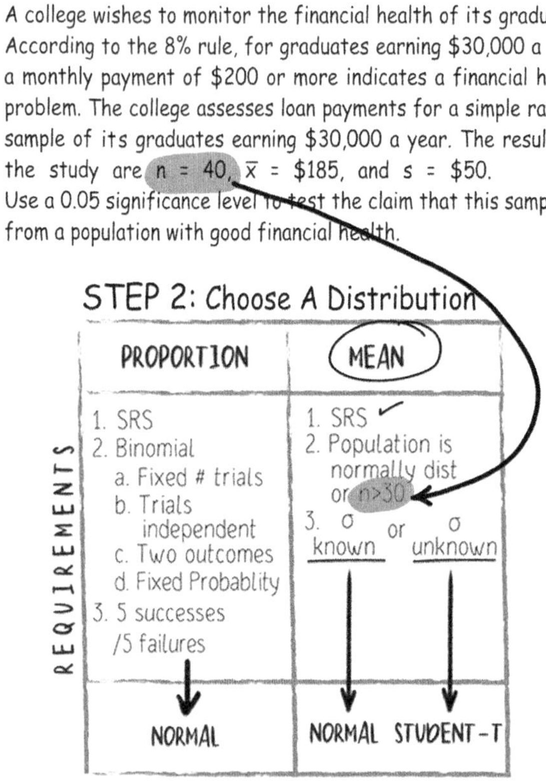

The problem statement states that n=40. Since the sample size is greater than 30, we can check off the second requirement.

Our third check is a crucial but often overlooked step. "Is σ known or unknown?" If σ is known, use the normal distribution. If σ is unknown, use the Student t distribution.

It's a sneaky step for two reasons. First, some texts just skip over the first scenario when σ is known, so when those questions pop up, students don't even realize they need to check this. Second, the sample standard deviation is always given in the problem statement, so some students assume that this means the standard deviation for the population is known. It's not! The sample standard deviation, s, is not the same as the population standard deviation, σ! Review your chart of parameters and statistics, then read the problem statement to determine if σ is known.

	population parameter	sample statistic
mean	μ	$\bar{x}$
proportion	p	$\hat{p}$
standard deviation	σ	s

A college wishes to monitor the financial health of its graduates. According to the 8% rule, for graduates earning $30,000 a year, a monthly payment of $200 or more indicates a financial health problem. The college assesses loan payments for a simple random sample of its graduates earning $30,000 a year. The results of the study are $n = 40$, $\bar{x} = \$185$, and $s = \$50$.
Use a 0.05 significance level to test the claim that this sample is from a population with good financial health.

Common Mistake: confusing s and σ
s = sample standard deviation
σ = population standard deviation

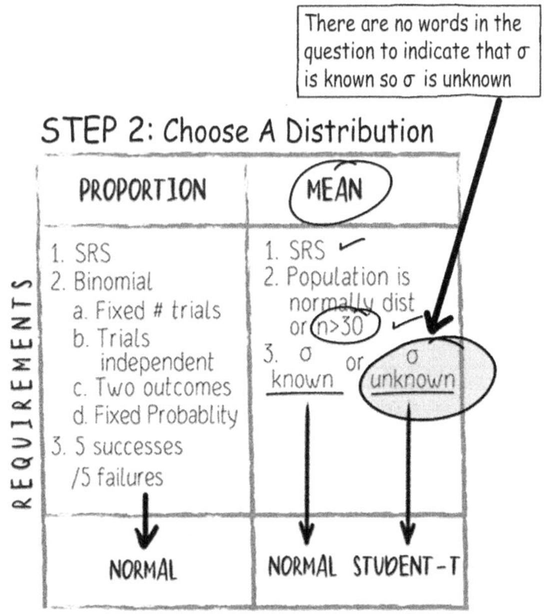

Make-Your-Life-Easier Tip: If your teacher and book skip over those rare questions where the population standard deviation is known, then requirement #3 will always be σ is unknown, and you will always use the Student t test. Many teachers simply mention that the rare scenario where σ is known exists, but don't include those questions in homework and tests. If that's your class, don't stress about requirement 3. σ will always be unknown.

Choose a Distribution

Once the requirements for a hypothesis test regarding a mean have been satisfied, follow the arrow in the chart.

The third step will determine which of the two distributions to use. In this problem only, the sample standard deviation is provided; therefore, the Student t distribution is the appropriate choice for the hypothesis test.

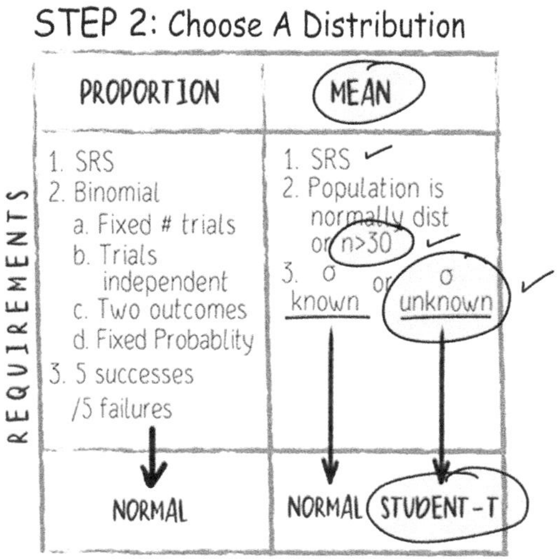

The hypothesis test performed will be the 1-sample test of means on the Student t distribution.

Common Question: What distribution do I use if both the population and sample standard deviations are known? If the population standard deviation is given, use the normal distribution. If both the population and sample standard deviations are given, then σ is known. Use the normal distribution. If only the sample standard deviation is given, then σ is unknown. Use the Student t distribution.

22. STEP 3 SKETCH THE CURVES

It's very easy to plug in a wrong number in the calculator, get a test statistic that makes no sense, and get the wrong answer for your hypothesis test. Sketching the curves will help you to catch those errors. Even if your teacher doesn't require you to show it, we insist you do anyway. Here are the two curves to fill in:

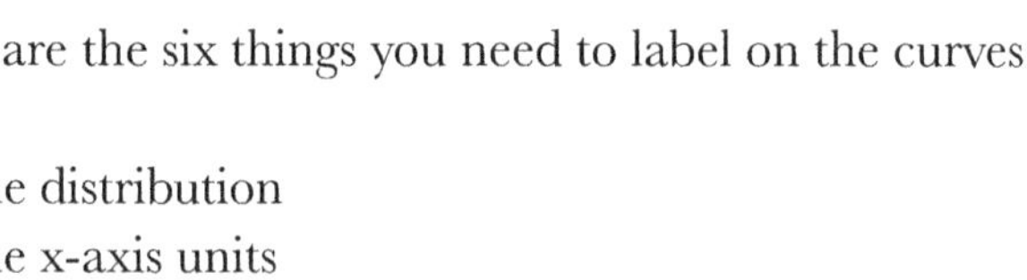

Here are the six things you need to label on the curves:

1. the distribution
2. the x-axis units
3. the mean of the curve
4. the standard deviation of the curve
5. the significance level area
6. the critical value(s)

Below is the checklist that we'll follow for our hypothesis test about a population mean. Notice that it's the same checklist we used when doing the hypothesis test about the population proportion.

	Sketching Curves Checklist		
	CHECKLIST	OBSERVATION SPACE	STANDARDIZED SPACE
1.	distribution ☐	normal / Student t	standard normal / standardized Student t
2.	units (x-axis) ☐		NO CHOICE standard deviations
3.	mean of curve ☐	μ = Null Hypothesis	NO CHOICE $\mu = 0$
4.	standard deviation of curve ☐	circle one $\sqrt{\frac{pq}{n}}, \frac{\sigma}{\sqrt{n}}, \frac{s}{\sqrt{n}}$	circle one 1 or ≈ 1
5.	significance level α (area) ☐	same area, both curves α = ____	Left Tailed Right Tailed Two Tailed
6.	critical value (x-axis) ☐	x_α, $x_{\alpha/2}$	z_α, $z_{\alpha/2}$, t_α, $t_{\alpha/2}$
1.	observation/ test statistic (x-axis) ☐	observation $\hat{p}$ or $\bar{x}$ =	test statistic z or t =
2.	p-value (area) ☐	same area, both curves p-value = ____	Left Tailed Right Tailed Two Tailed

Remember that active learning is better than passive learning. This means that you'll learn more if you try to solve the question yourself rather than simply reading about how to answer the question. Solving a problem versus reading about how to solve a problem is similar to the difference between jumping in the pool to swim and reading about swimming. If you haven't already, try to graph these curves yourself, then check your work. Reviewing the details below will be more meaningful if you've already worked through the problem yourself.

One Final Reminder: Drawing the curves is a crucial step to help you make sense of your work, understand the problem, and provide a common-sense check for when errors occur. Do not skip this!!

Label the Distribution and Units

First, label the distribution. We're using the Student t distribution! We just checked the requirements in the last step. Easy peasy.

Second, label the x-axis units. Here, the x-axis label is dollars. If we look back to Step 1, State the Hypotheses, we see that our original claim was "All graduates earning \$30,000/year have an average monthly payment of less than \$200." What are the units for \$200? Dollars!

STEP 3	OBSERVATION SPACE	STANDARDIZED SPACE
distribution	normal / (Student t)	standard normal / (standardized Student t)
x-axis units/ label	dollars	NO CHOICE standard deviations
mean of curve	μ = Null Hypothesis $\mu = 200$	NO CHOICE $\mu = 0$
standard deviation of curve	$\frac{s}{\sqrt{n}} = \frac{50}{\sqrt{40}} \approx 7.9$	circle one 1 or ($\approx$ 1)

Label the Mean of the Curve

What is the mean of the curve? We use the null hypothesis, $\mu = 200$, to label the mean of the curve.

STEP 3	OBSERVATION SPACE	STANDARDIZED SPACE
distribution	normal / (Student t)	standard normal / (standardized Student t)
x-axis units/ label	dollars	NO CHOICE standard deviations
mean of curve	μ = Null Hypothesis $\mu = 200$	NO CHOICE $\mu = 0$
standard deviation of curve	$\frac{s}{\sqrt{n}} = \frac{50}{\sqrt{40}} \approx 7.9$	circle one 1 or ($\approx$ 1)

Here's where the mean is placed on the curve.

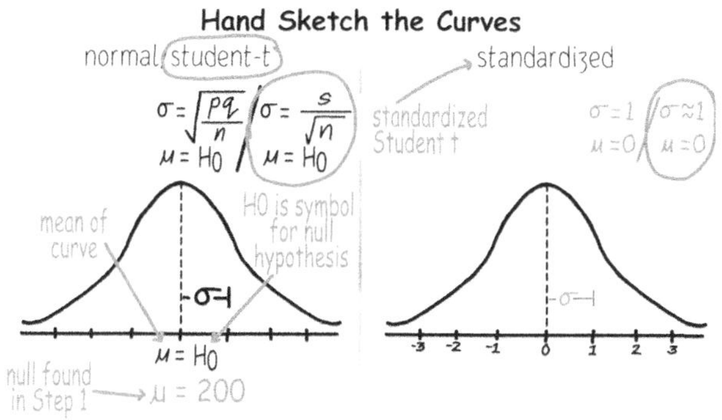

Common Question: When drawing the curve, do I use the population mean or the sample mean? The mean of the curve is the null hypothesis, $\mu = 200$, which is the population mean. Do *not* use \$185 as the mean of the curve. That's the mean for the sample.

Common Mistake: Mixing up parameters and statistics. The population mean, which is what's depicted on the curve, is always given by the null hypothesis because the null hypothesis is the assumed population mean. The sample mean will be used as the observation. We'll cover more on that when we show you the Critical Value Test.

So often, in the middle of a long problem like this, students get mixed up, and keeping track of what is a parameter and what is a statistic suddenly gets tricky. In case that's you, here's a quick recap of the problem.

GATHER INFORMATION FOR A PROBLEM INVOLVING A MEAN

	parameter	statistic
mean	Null Hypothesis $\mu = 200$ dollars	$\bar{x} = 185.00$ dollars
standard deviation	σ = UNKNOWN	$s = 50$ dollars
size	N = UNKNOWN	$n = 40$

Label Standard Deviation of the Curve

The formula for the standard deviation of the curve depends on our population parameter and distribution. In this problem, we are using the Student t distribution for a hypothesis test about a population mean.

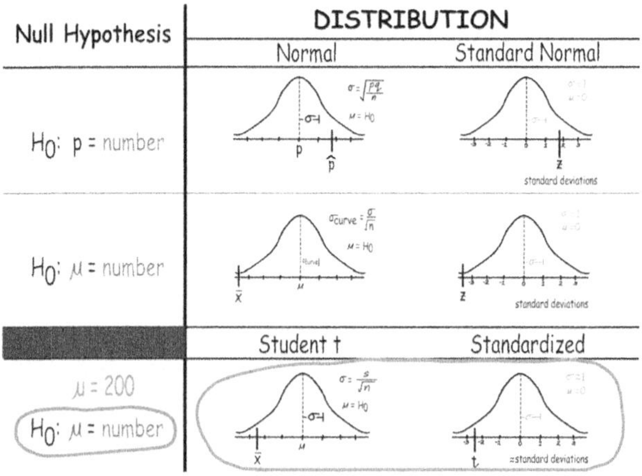

Because σ is unknown, we will calculate an approximation of the standard deviation using s and n. Therefore, the standard deviation of our curve with s = 50 and n = 40:

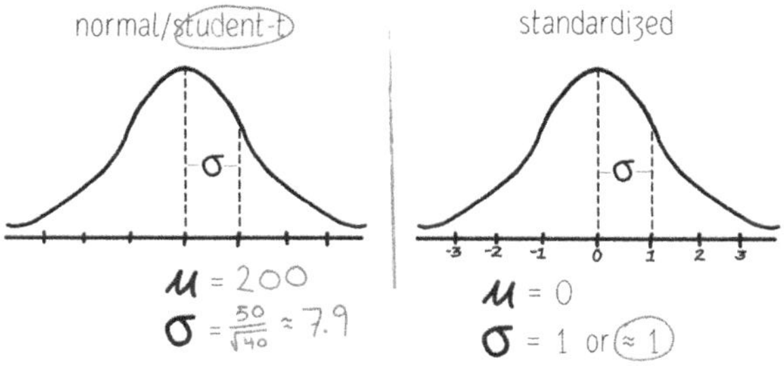

Common Question: Why do we divide by the square root of the sample size? Forgetting that our curve is a sampling distribution is a common mistake. When estimating the standard error for the mean, the larger the sample size, the smaller the standard deviation. You account for this by dividing s by the square root of the sample size. For more on this, see the sampling distributions section earlier in the book.

Here's our checklist updated with the standard deviation.

STEP 3	OBSERVATION SPACE	STANDARDIZED SPACE
distribution	normal / (Student t)	standard normal / (standardized Student t)
x-axis units/ label	dollars	NO CHOICE standard deviations
mean of curve	μ = Null Hypothesis μ = 200	NO CHOICE **μ = 0**
standard deviation of curve	$\frac{s}{\sqrt{n}} = \frac{50}{\sqrt{40}} \approx 7.9$	circle one 1 or (≈ 1)

Here's what our curve looks like so far.

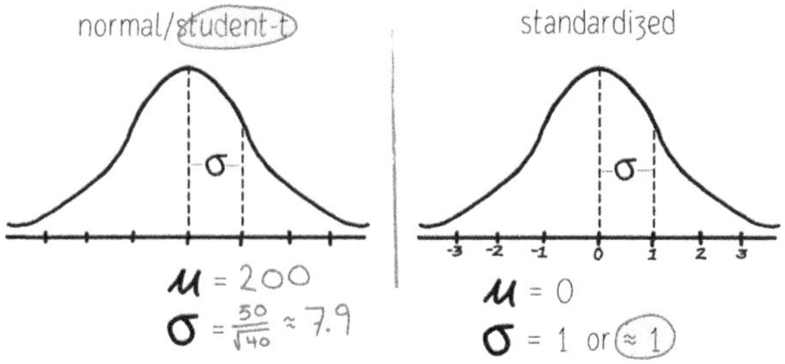

Common Question: Why is the standardized Student t distribution approximately one? Shouldn't it be greater than one? Yes! When we draw our pictures with $\sigma = 1$, this is slightly incorrect. The standardized Student t distribution has a standard deviation close to one but a bit larger than one. The picture we draw is not intended to be a mathematically perfect representation of the data; the picture of both curves should be considered a quick sketch that helps us make sense of the information we have gathered and check our work. While imperfect, the curve sketches we construct here are useful.

Next, we use the standard deviation of the curve to calculate the labels on the x-axis. We label three standard deviations above and below the mean. From left to right on the x-axis, three standard deviations above and below the mean are:

$$\mu - 3\sigma = 200 - 3 * 7.9 = 176.3$$
$$\mu - 2\sigma = 200 - 2 * 7.9 = 184.2$$
$$\mu - \sigma = 200 - 1 * 7.9 = 192.1$$
$$\mu = 200$$
$$\mu + \sigma = 200 + 1 * 7.9 = 207.9$$
$$\mu + 2\sigma = 200 + 2 * 7.9 = 215.8$$
$$\mu + 3\sigma = 200 + 3 * 7.9 = 223.7$$

Here's where the values go on the x-axis.

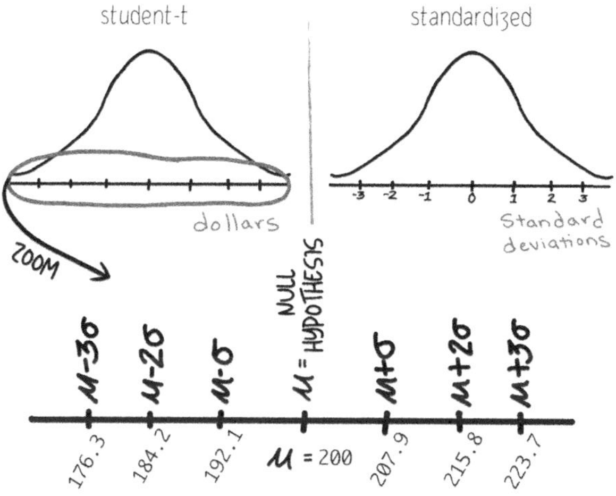

Shade the Significance Level Area

Our alternative hypothesis from Step 1, State the Hypotheses, determines the number of tails in our hypothesis test and where we shade the significance area.

Given our alternative hypothesis, $\mu < 200$, we have a one-tailed, left-tailed test.

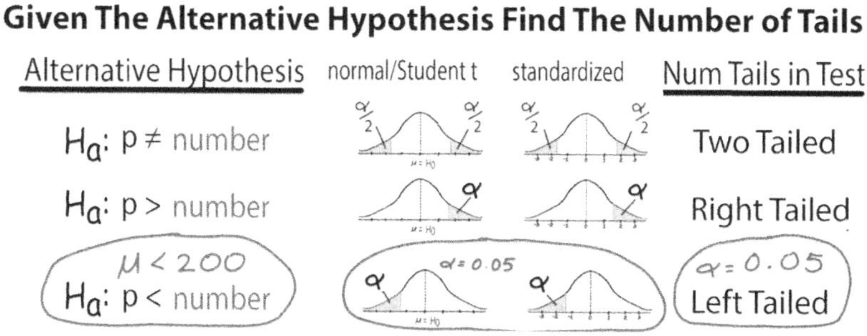

A one-tailed test tells us to place the entire significance level area in the left tail; this area will correspond with a single critical value on the x-axis.

Adding the significance level area **α** to our graph, we get:

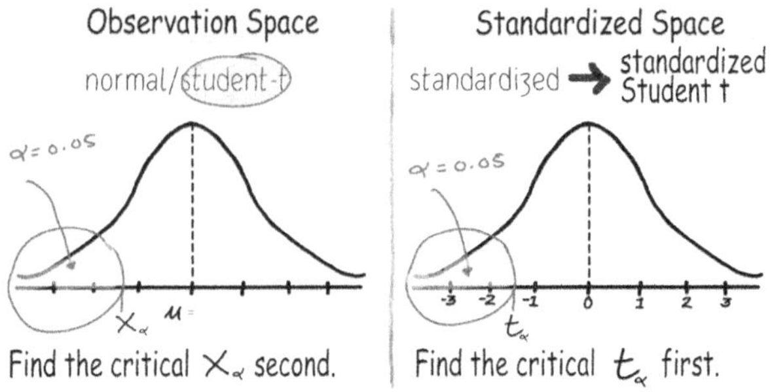

Label the Critical Value and Rejection Region

In the last section, we *generally* shaded the significance level area, but we didn't know where the exact border for the shaded area was. Now, our job is to make it exact by finding the critical value:

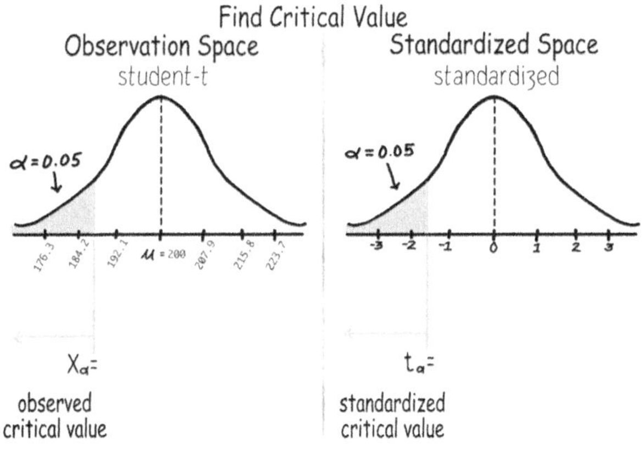

To find the critical value on the standardized curve (right graph), you can use the TI-84 calculator. The Student t distribution is defined by degrees of freedom, where $df = n\text{-}1$, so we use the invT(area, df) function. It's important to note that the invT(area, df) function on your calculator requires your input to be the area *to the left* of the critical value.

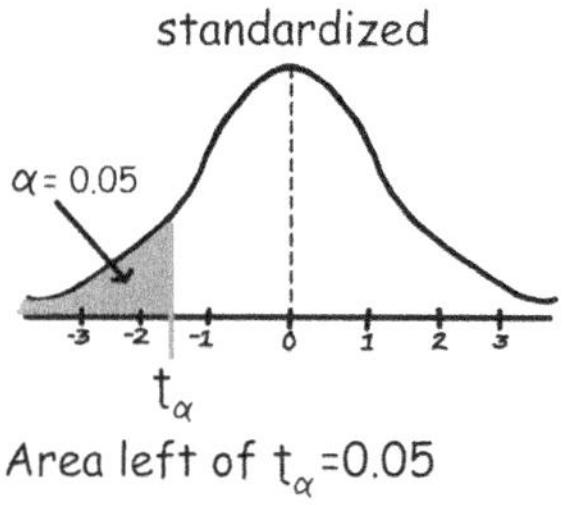

Find standardized critical value on the standardized Student t: t_α
Calculator inputs for function invT(area, df)

area = area to the left of the critical value,
a decimal between 0 < area < 1
df = degrees of freedom,
in this case - the sample size minus one, n-1

To find the standardized critical value (on the right graph), use the function invT(0.05, 39) = -1.684875066.

Here are the step-by-step instructions:

1) Choose 2nd, then Distr

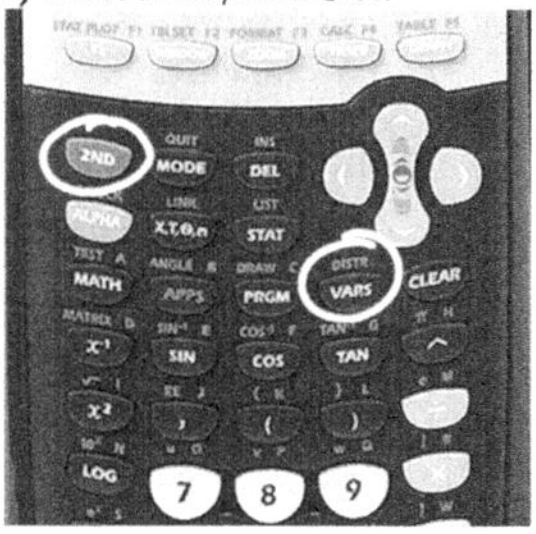

2) Choose 4:invT

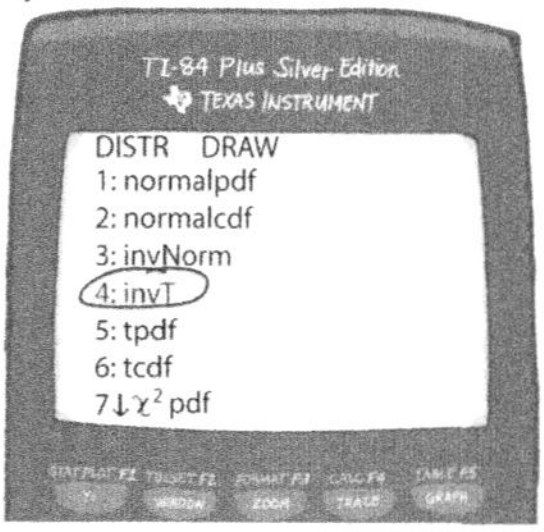

3) Add inputs

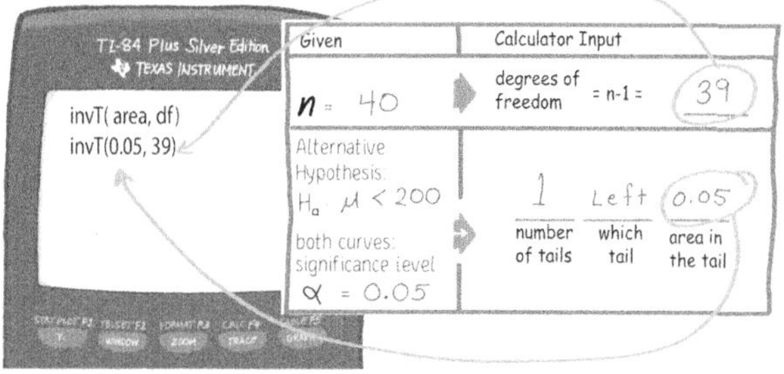

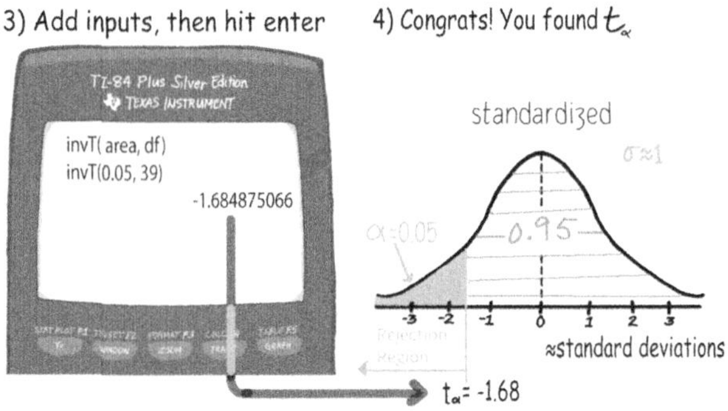

When you label this critical value on your graph, you'll round the numeric output to two decimal places.

Next, you need to find the critical value on the Student t curve by hand. Yes, unfortunately, the critical value on the Student t curve (left graph) must be done by hand. Use algebra to solve the equation for the variable you wish to find; solve the equation for the observed critical value.

observation = standard score * (standard deviation) + mean

Plugging in the variables for the Student t distribution, this equation becomes:

equation in variables

$$X_\alpha = t_\alpha * \left(\frac{s}{\sqrt{n}}\right) + \mu$$

Next, replace the variables with numeric values.

$$\begin{aligned} X_\alpha &= t_\alpha * \left(\frac{s}{\sqrt{n}}\right) + \mu \\ &= -1.68487 * \left(\frac{50}{\sqrt{40}}\right) + 200 \\ &= 186.67993 \end{aligned}$$

The units of measure for the observation are dollars, so we round the final answer to the nearest cent, \$186.68.

Common Question: What happens if I round to fewer decimal places? When you're plugging in and working through problems, keep at least five decimal places. Rounding to fewer decimal places can lead to slightly different outputs, which your teacher or online homework tool may mark as incorrect.

Critical Value

t_α = -1.68 Rounded

t_α = -1.68487 Truncated at 5 decimals

If the rounded values above are plugged in here, you likely will not obtain the correct answer!

Lastly, we use the critical values to label the rejection region.

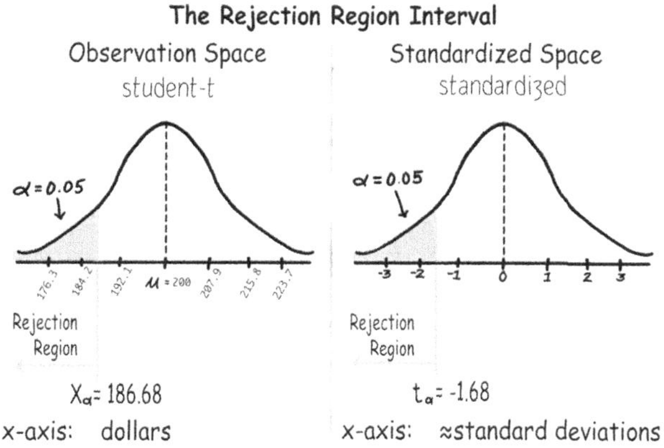

Your teacher may ask you to write out the rejection region as an inequality or in interval notation. In this case, in the observation space (left graph), the rejection region is:

- Inequality: $x \leq 186.68$
- Interval Notation: $(-\infty, 186.68]$

In the standardized space (right graph), the rejection region is:

- Inequality: $x \leq -1.68$
- Interval Notation $(-\infty, -1.68]$

What does the rejection region represent? In Step 4, Do the Test, if an observation or test statistic falls in the rejection region, the null hypothesis is rejected. The rejection region is the criteria we use in the Critical Value Test to decide if we will reject or fail to reject the null hypothesis. Similarly, the shaded significance level area is the criteria we use in the P-Value Test to determine if we reject or fail to reject the null hypothesis! Our entire setup for our hypothesis test is done. All we need now is the test statistic and p-value.

23. STEP 4 DO THE TEST

The goal here is to execute the hypothesis test. At the end of Step 4, Do the Test, we will either reject the null or fail to reject the null hypothesis.

These are the Step 4 building blocks:

1. Calculate the p-value and test statistic.
2. Complete the curve sketch.
3. Do the P-Value Test.
4. Do the Critical Value Test.

Calculate the P-Value and Test Statistic

First, find the observation for the Student t curve (left graph). In the observation space, the observed test statistic is represented as a sample statistic. Pull out a chart of parameters and statistics and figure out what variable represents the observed sample statistic. Read the problem statement, and underline the observed sample statistic.

Out of all graduates earning \$30,000 a year, a college collects data in a simple random sample of 40 such graduates. Here are the results: $n=40$, $\bar{x}$ = \$185, s = \$50. Use a .05 significance level to test the claim that all graduates earning \$30,000 a year have an average monthly payment that is significantly lower than \$200.

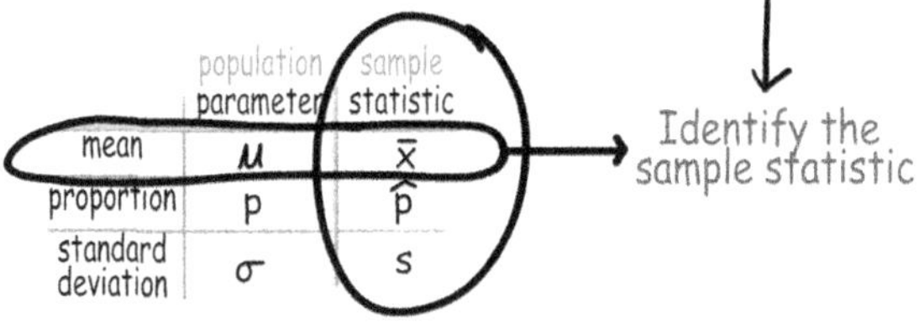

The observed sample statistic is $\bar{x}$ = \$185.

Next, find the test statistic for the standard curve (right graph).

The simplest way to find the test statistic is to use the TI-84 calculator. To calculate the test statistic, follow these steps:

● Choose Stat

● Choose Test, then 2: T-Test

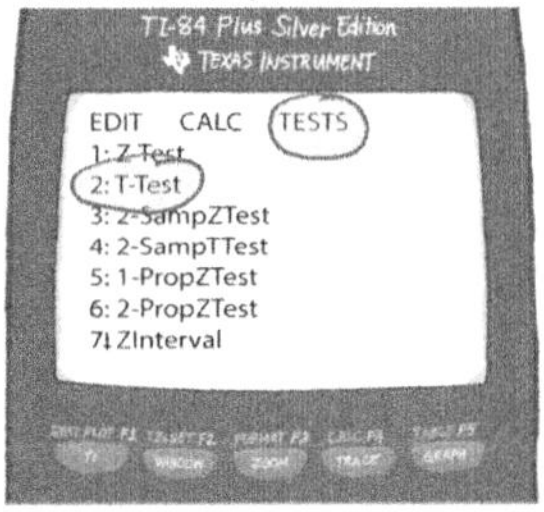

Ti-84 Calculator: T-Test INPUTS

Observation Space

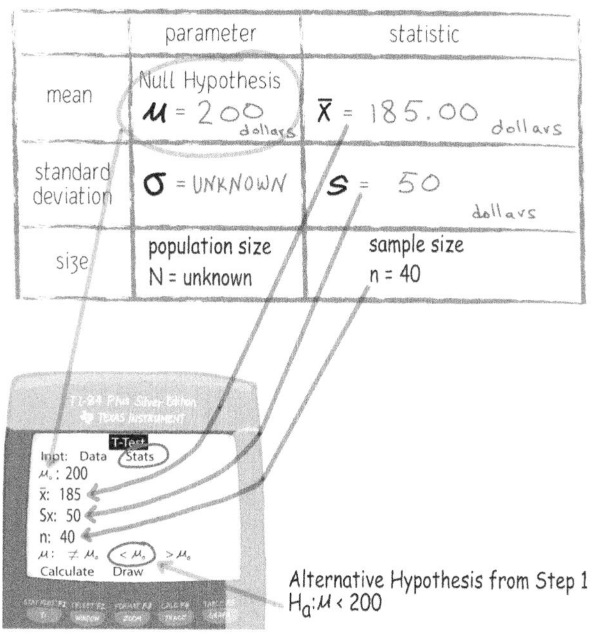

Ti-84 Calculator: T-Test OUTPUTS

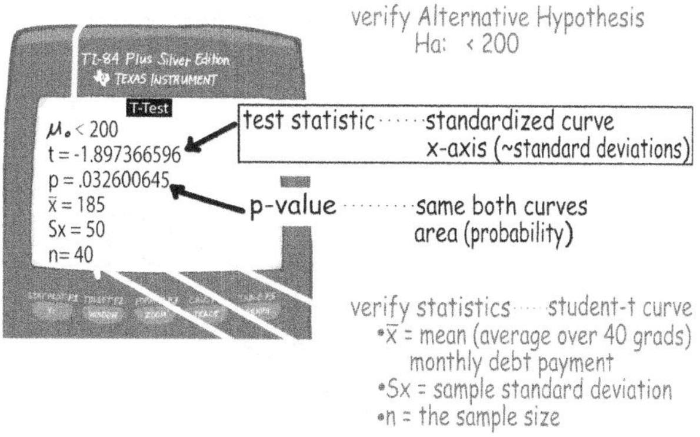

There are a number of outputs listed on the screen. The test statistic is t = -1.897366595.

Complete the Curve Sketch

Add the observation and the test statistic to the appropriate curve.

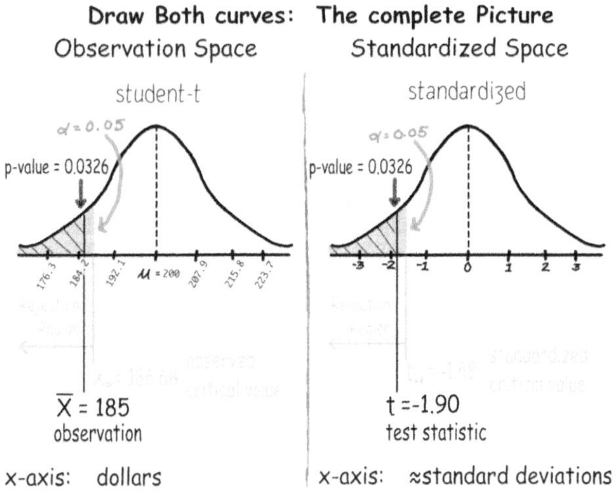

Do the P-Value Test

The P-Value Test compares two areas; that comparison of two areas tells us whether or not the test statistic is significant. If the area in the tail of the test statistic is smaller than the preselected significance level, the test statistic is significant. When we say it is significant, we mean that the test statistic is so unusual that we suspect it was not generated by a population with the mean used to create the curve in the observation space.

Return to the TI-84 calculator T-Test. Find the function output that is the p-value.

Ti-84 Calculator: T-Test OUTPUTS

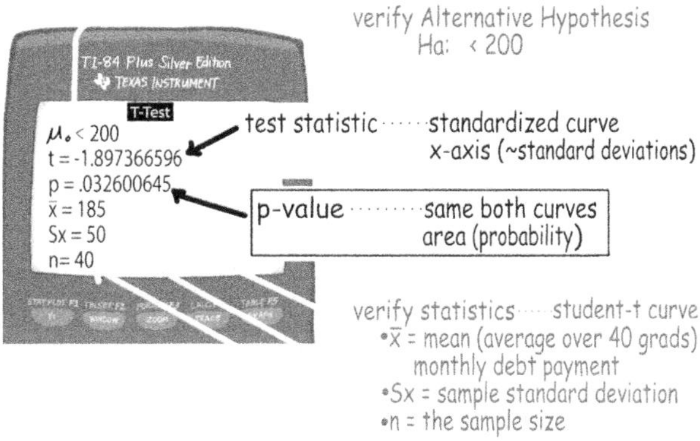

The p-value is 0.0326. This is the area in the left tail. On the Student t curve, this is the area above the x-axis from (-∞, observation]; on the standardized Student t curve, this is the area above the x-axis from (-∞, test statistic]. It does not matter whether we are in the observation space or the standardized space; in both spaces, the area in the tail is the same size.

P-Value Method: Compare Areas

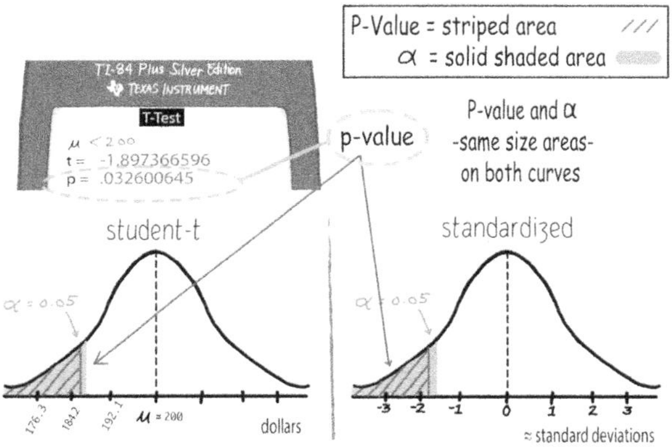

Sketching Curves Checklist

	CHECKLIST		OBSERVATION SPACE	STANDARDIZED SPACE
5.	significance level α (area)	☑	same area, both curves α = 0.05	(Left Tailed) Right Tailed Two Tailed
6.	critical value (x-axis)	☑	X_α = 186.68	t_α = -1.68
1.	observation/ test statistic (x-axis)	☑	observation $\bar{X}$ = 185.00	test statistic t = -1.90
2.	p-value (area)	☑	same area, both curves p-value = 0.0326	(Left Tailed) Right Tailed Two Tailed

Next, compare the p-value area to the significance level area. Look at the complete picture of both curves. Note that the striped p-value area is smaller than the solid-colored significance level area. Sometimes, it's easier to see this in the curve sketch; other times, it's easiest to just compare the numeric values in the chart.

Another way to show what we see in the graphs is:

- If p-value < $\boldsymbol{\alpha}$, we reject the null hypothesis.
- If p-value $\geq$ $\boldsymbol{\alpha}$, we fail to reject the null hypothesis.

In our case, p = 0.032 < .05, so we reject the null hypothesis.

Helpful Tip: A cheesy, albeit very effective, way of remembering this rule is: When the p-value is low, the null must go. In this problem, we have .032, which is low (i.e., lower than .05), so the null must go, meaning we reject the null hypothesis, H_0.

TEST	P-VALUE METHOD	
CALCULATE	P-Value = 0.0326	
TEST	Is P-Value ≤ α ? 0.0326 ≤ 0.05	
	No	Yes
DECISION	Fail to Reject H_0	Reject H_0

Since the p-value, 0.0326, is smaller than the significance level, 0.05, the conclusion is that we reject H_0. The p-value is smaller, so we reject the null hypothesis.

Do the Critical Value Test

The Critical Value Test compares the critical value and test statistic on the x-axis. If the test statistic falls in the rejection region interval, then we reject the null hypothesis, H_0.

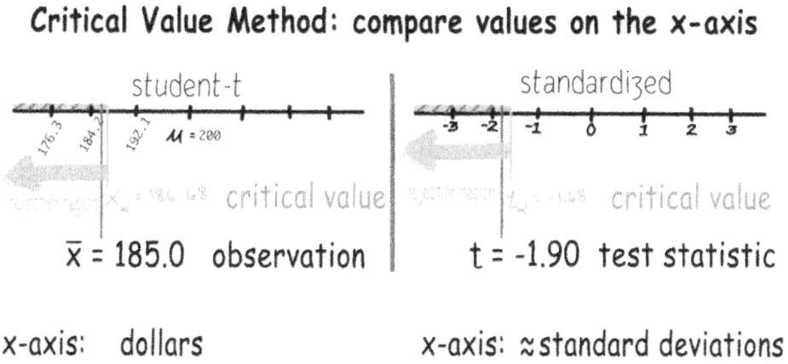

Let's look at the close-up of the axes. On the Student t curve (on the left), we see that the observation, 185, is to the left of the critical value x_α = 186.68. The rejection region interval is (-∞,

186.68], and the observation of 185.00 falls inside the rejection region interval. This tells us to reject the null hypothesis, H_0.

On the standardized Student t curve (on the right), we see that the test statistic t = -1.90 is to the left of the standardized critical value t_**α** = -1.68. The rejection region interval is (-∞, -1.68], and the test statistic of -1.90 falls inside the interval. This tells us to reject the null hypothesis, H_0.

Notice that both graphs, the one in the observation space and the one in the standardized space, lead to the same conclusion. This must always be the case.

Common Mistake: Thinking that, if the test statistic falls into the rejection region, you will reject your *claim*. That's not the case. The test is not about the *claim*; it is about the *null hypothesis*. The claim and the null hypothesis may be the same, but they can also be different. The test determines if we reject or fail to reject the null hypothesis.

Common Question: Notice that the *observation and test statistic are both in the rejection region*, the part of the x-axis to the left of the critical value. At this point, many students ask, "Why didn't my teacher just say that?" My answer is that your teacher probably did, but they might have used different words to say it. One teacher might say, "*The test statistic is closer to the tail than the critical value*." Another teacher might tell you, "*The observation is more extreme than the critical value*." Those are simply different ways of saying that the observation and test statistic fall within the rejection region.

Full Step 4 Solution

Sketching Curves Checklist

	CHECKLIST	OBSERVATION SPACE	STANDARDIZED SPACE
1.	distribution ☑	normal / (Student t)	standard normal / (standardized Student t)
2.	units (x-axis) ☑	dollars	NO CHOICE standard deviations
3.	mean of curve ☑	μ = Null Hypothesis $\mu = 200$	NO CHOICE $\mu = 0$
4.	standard deviation of curve ☑	circle one $\frac{s}{\sqrt{n}} = \frac{50}{\sqrt{40}} \approx 7.9$	circle one 1 or ($\approx$ 1)
5.	significance level α (area) ☑	same area, both curves α = 0.05	(Left Tailed) Right Tailed Two Tailed
6.	critical value (x-axis) ☑	X_α = 186.68	t_α = -1.68
1.	observation/ test statistic (x-axis) ☑	observation $\bar{X}$ = 185.00	test statistic t = -1.90
2.	p-value (area) ☑	same area, both curves p-value = 0.0326	(Left Tailed) Right Tailed Two Tailed

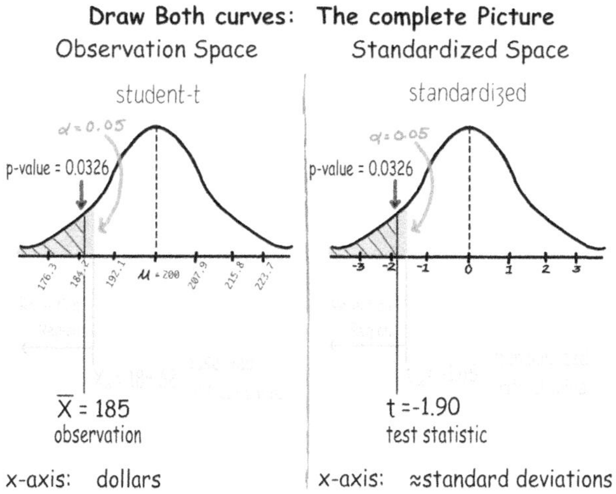

Filling in our P-Value Test and Critical Value Test results from earlier, we see that both tests reach the same conclusion.

TEST	P-VALUE METHOD	CRITICAL VALUE METHOD
CALCULATE	P-Value = 0.0326	Test Statistic = -1.90
TEST	Is P-Value ≤ α? $0.0326 \leq 0.05$ No / Yes (circled)	Is Test Statistic in the Rejection Region? No / Yes (circled)
DECISION	Fail to Reject H_0 / Reject H_0 (circled)	Fail to Reject H_0 / Reject H_0 (circled)

According to the Critical Value Test, we should reject the null hypothesis H_0. The P-Value Test also indicates that we should reject H_0. This is good because, even though the different methods compare different things, both methods use the same graph and should give us the same answers.

Now, let's answer some of the questions you might have:

Common Question: If the standard deviation of the standardized Student t curve is larger than 1, how do I draw a sketch of the curve? Regardless of the precise value, the standard deviation of the standardized Student t will always be close to 1. Draw the sketch by hand as if the standard deviation was precisely 1. This imperfect sketch is good enough and will help you understand what is going on in the problem.

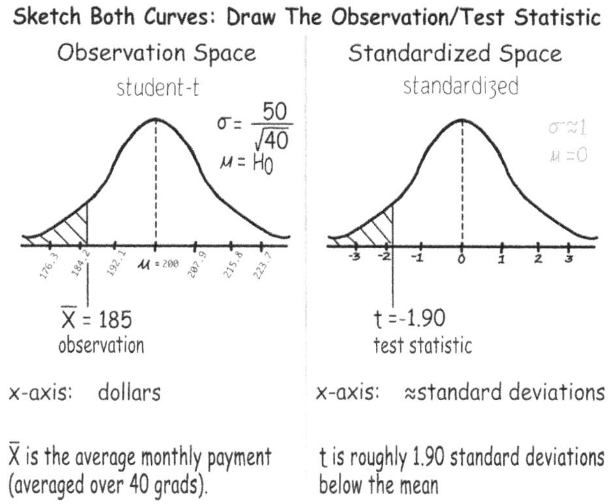

If you later graph a perfect picture using technology, it will look similar, and this hand sketch will allow you to verify that your computer-generated graphs make sense.

Common Question: For the test statistic, how many places should I round to? In general, unless otherwise specified, test statistics are rounded to two decimal places. We obtained a calculator output of t = -1.897366956; the rounded test statistic is t = -1.90. That being said, your instructor or text may specify any level of rounding and you should follow their directions.

Common Question: What does the test statistic tell us about the location of the observation? The test statistic is 1.90 standard deviations below the mean.

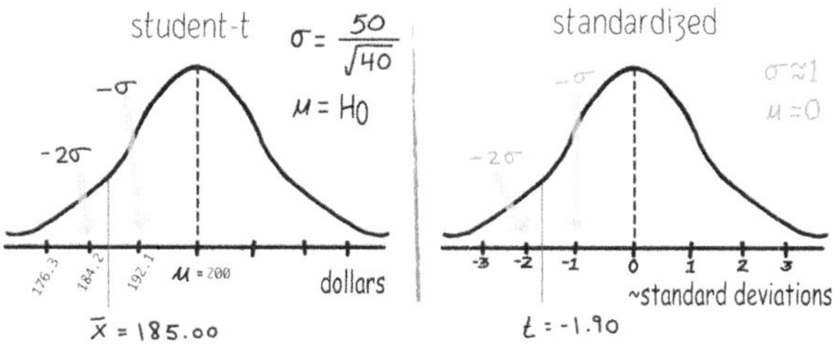

Notice that both the observation and test statistic are 1.90 standard deviations below the mean; in terms of standard deviations, both are in the same location. Once you see the observation on the x-axis of the Student t curve, you can predict roughly where the test statistic will be on the standardized curve. Since the observation falls between the first and second standard deviation below the mean on the Student t curve, you know that the test statistic must also be between the first and second standard deviation below the mean on the standardized curve. When you evaluate the test statistic, check the numeric solution for errors by making sure the observation and test statistic are in the expected matching locations.

Remember, if the observation is on a Student t curve, the test statistic on the standardized curve is called a t-score. Given the mean, standard deviation, and observation from the Student t curve, you can find the test statistic.

Common Question: How do I calculate the test statistic by hand? This answer is long, but if your teacher asks you to calculate the test statistic by hand, here's how to do it. You'll use the same z-score equation we used for the 1-Sample Proportions. Here it is, written in its general form:

equation in words

$$\frac{\text{observation - mean}}{\text{standard deviation}} = \text{standard score}$$

Second, visualize what it means to find the test statistic. Given the observation, find the test statistic. Another way to visualize this concept is to think of a function; that function will input the observation and output the test statistic. Draw a picture of what is known and then indicate what needs to be found.

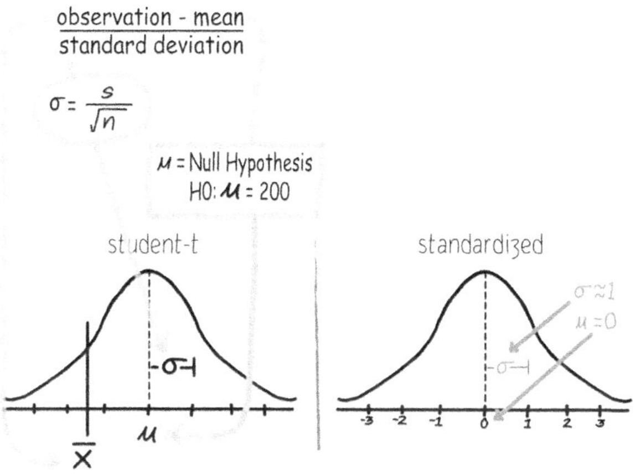

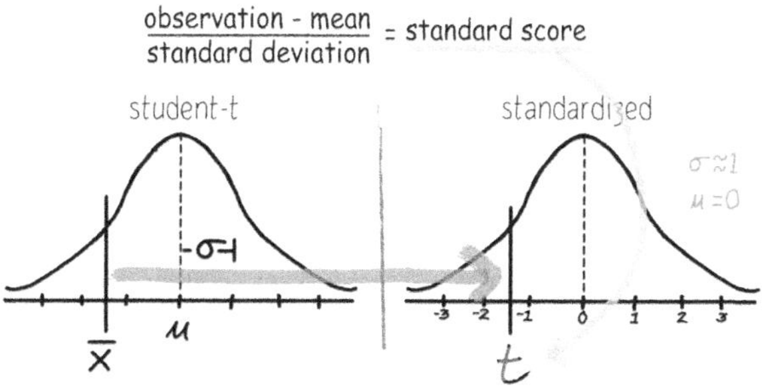

Third, gather the information you have in a table.

STEP 3	normal/student-t	standardized
x-axis units/ label	dollars	NO CHOICE standard deviations
mean of curve	μ = Null Hypothesis $\mu = 200$	NO CHOICE $\mu = 0$
standard deviation of curve	circle one $\frac{s}{\sqrt{n}} = \frac{50}{\sqrt{40}} \approx 7.9$	circle one 1 or ≈ 1
both curves: significance level $\alpha = 0.05$	Left Right Both	
critical value	$X_\alpha = 186.68$	$t_\alpha = -1.68$
STEP 4 observation/ test statistic	observation $\bar{X} = 185.00$	test statistic $t =$ Find by hand and by Calculator

equation in words

$$\frac{\text{observation - mean}}{\text{standard deviation}} = \text{standard score}$$

The information in the table can be used to rewrite the transformation equation from words to variables. Replace the words with the corresponding variables found in the table.

equation in variables

$$\frac{\bar{X} - \mu}{\frac{s}{\sqrt{n}}} = t$$

Finally, replace the variables with numeric values and evaluate the t-score. Remember, don't truncate numbers when you can input the exact amount; if you do truncate, carry five or more decimal places.

$$\frac{\bar{x} - \mu}{\frac{s}{\sqrt{n}}} = t$$

$$\frac{185 - 200}{\frac{50}{\sqrt{40}}} = t$$

$$-1.89736 = t$$

That matches the calculator value!

Common Question: How many decimal places should I use when performing a calculation? Rounding is always done at the end of a calculation. When performing a calculation, use all of the decimals that you have. For example, when calculating the test statistic by hand, do not round the standard deviation; input the entire fraction.

equation in variables

$$\frac{\bar{x} - \mu}{\frac{s}{\sqrt{n}}} = t$$

$$\frac{\bar{x} - \mu}{\frac{s}{\sqrt{n}}} = t$$

$$\frac{185 - 200}{\frac{50}{\sqrt{40}}} = t$$

$$-1.89736 = t$$

If you come across a situation in which you cannot input the standard deviation as a fraction, carry *all* the decimal places in your calculation.

$$\frac{185-200}{\frac{50}{\sqrt{40}}} = t$$

$$\frac{185-200}{7.90569415} = t$$

$$-1.89736 = t$$

Then round the final answer to the appropriate number of decimal places. In this case, the test statistic t = -1.90.

Common Question: How do you find the p-value using the test statistic and the tcdf() function? Some teachers require you to find the p-value using the test statistic. To do that, you'll have to use the information from the standardized space. The TI-84 calculator can only calculate the p-value for the standardized Student t function from the standardized space.

You'll need to input degrees of freedom and the upper and lower bounds from the standardized space into the tcdf function.

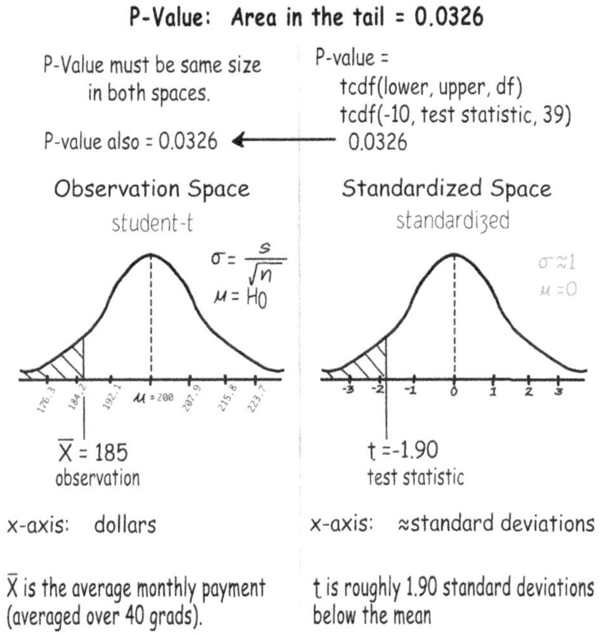

Common Question: What should the upper and lower bounds be? First, it is important to understand that the tcdf function only allows inputs from the standardized space. To find the correct lower and upper bounds, you must be looking at the standardized curve. The upper and lower bounds are values on the x-axis, as shown below.

The upper bound is the test statistic. The lower bound is technically negative infinity since the axis goes on forever. The TI-84 calculator does not have a symbol for infinity; we substitute a large negative number. Here, we use -10 for the lower bound.

1) Choose 2nd, then Distr

2) Choose 6:tcdf(

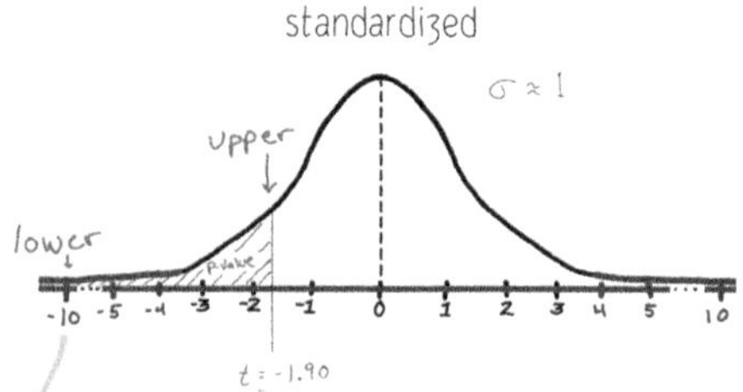

3) Add inputs, then hit enter

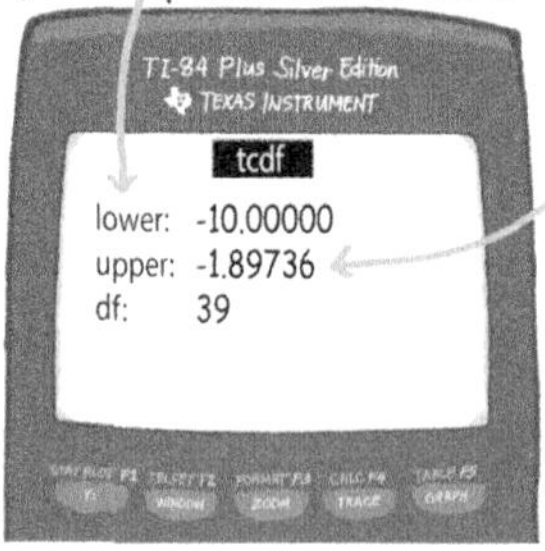

Test Statistic	
t = -1.90	Rounded
t = -1.89736	Truncated at 5 decimals

Use 5 decimal places for all inputs because you need the output to be accurate to 4 decimal places.

4) Interpret Outputs

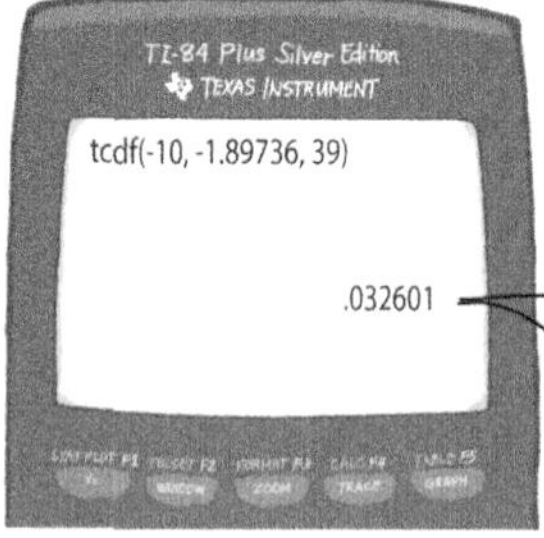

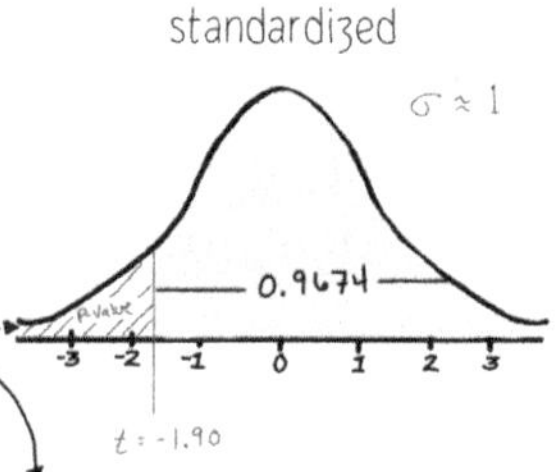

P-Value = 0.0326

The p-value is the area in the tail of the test statistic; it is a probability.

Common Question: Will I get a different answer if I choose a lower bound different than -10? As long as you choose a sufficiently large negative number, it doesn't make a difference. Below, we do the same calculation but with -100 for the lower bound.

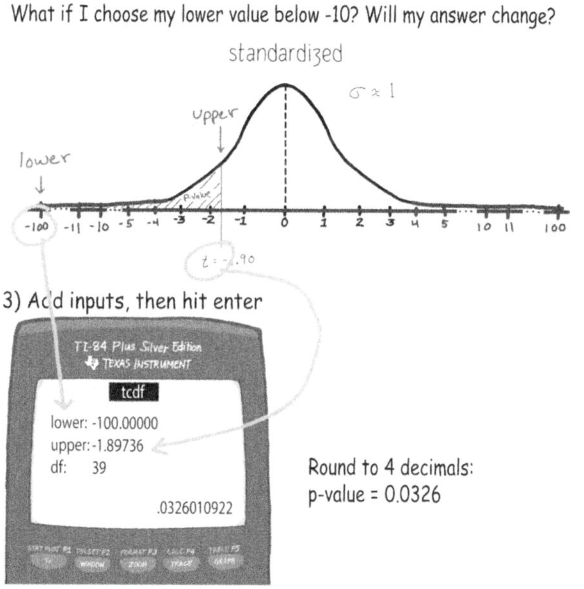

The area in the tail between -100 and -10 is so small it is approximately zero, so our answer does not change. If you are unsure which lower value to use, choose the smaller one.

The area in the tail between -100 and -10 is so small, it is approximately zero, so our answer does not change. If you are unsure which lower value to use, start by asking if it's a left tail or a right tail? If the desired area is in the left tail, choose the lower bound to be the number furthest to the left on the number line; choose the smallest number. In this case, -100.

24. STEP 5 STATE THE CONCLUSION

One of the online tools that some schools use for homework automatically grades hypothesis questions by this rubric:

- Null and alternative hypothesis: 1 point
- P-value: 1 point
- Conclusion: 1 point

Some teachers grade the same way. I tell you this in case you're one of the many students who is intimidated by writing the conclusion of a hypothesis question. Do not skip this! Do not do all the work correctly only to get ⅔ of the points because you're too afraid to do the difficult work of figuring out how to write the final conclusion. You've made it this far. Don't give up now.

Here's how Step 5 breaks down:

1. Follow the flow chart.
2. Write the basic conclusion.
3. Write the meaningful conclusion.

Try to do Step 5, State the Conclusion, yourself first. When you're done, check the full answer on the next page.

Follow the Flow Chart

Your first step here is to fill in your handy flow chart. Circle Step 1, Circle Step 4, and the initial wording for your conclusion is right there for you!

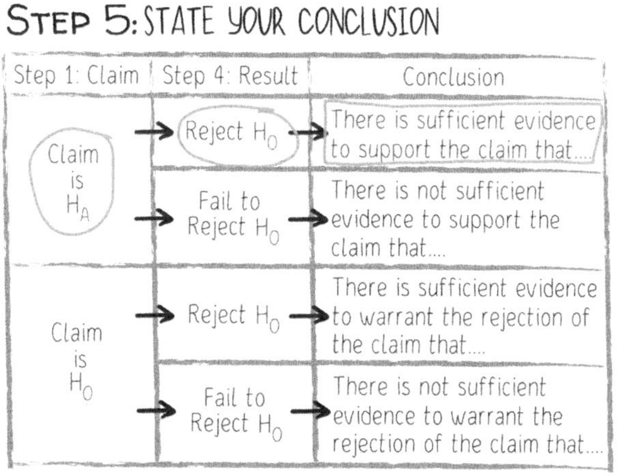

First, refer back to Step 1 to see if our original claim is H_0 or H_a. Ours was H_a.

Second, look back to Step 4 to see if your decision was to reject or fail to reject H_0. Ours was to reject H_0.

Write the Basic Conclusion

We combine the wording above with the wording from our original claim in Step 1 to get the following.

There is sufficient evidence to support the claim that recent graduates come from a population with a mean monthly payment of less than \$200, which is considered too low relative to their \$30,000 a year salary.

Is that answer the boring and cookie-cutter answer? Does it suck the creative and analytical thinking out of the last step of the test? Absolutely! Then why did we include it? Because it cuts down on the confusion.

The very first time I taught Statistics, I asked students to just "state the conclusion." Do you know what happened? *Everyone* got lost. Some would forget what their original claim was. Some would mistakenly think the null hypothesis was the claim. Some got it right but got confused on the wording. There were so many different ways students got confused that we decided to streamline this step:

- Go back to Step 1 and to see if your claim is H_0 or H_a.
- Go back to Step 4 to see if your decision was to reject or fail to reject H_0.
- Follow the flow chart to see what wording you should use for your conclusion.

If you find yourself getting confused about the wording, use the Step 5 flow chart until you can figure out the wording for yourself. Just as we used a multiplication table to check ourselves when we were learning multiplication, it's okay to use the chart above to check yourself as you're learning hypothesis testing.

Write the Meaningful Conclusion

If your teacher is one of those (awesome) teachers who wants you to know what that means, then you may also need to go a step further and explain the significance of the sentence you just wrote.

Does the result suggest that each of the forty graduates has a monthly payment that is too low? No, there is significant evidence the sample is from a population with an average monthly loan payment that is above \$200, but this is a sample distribution of sample means, not individual values, so it is unclear how many of the individuals in the sample have loan payments that put them in a difficult financial situation. The average payment across all forty is too high, indicating that one or more has poor financial health and is at increased risk for default.

A+ STUDENT QUESTIONS

Throughout the book, there were *A+ Student Questions* that required answers too long to include in the book. These included:

- Where do the transformations between normal curves and standard normal curves come from?
- When talking about degrees of freedom, what do you mean by the number of values that are free to vary?
- How do sampling distribution of sample proportions (SDSP) and the CLT relate to sample proportions?
- What's the difference between significance levels and confidence levels?
- Why do we double or half the p-value in two-tailed tests and how do we know which to do?
- Why is the mean and standard deviation of data different from the mean and standard deviation of the curve?
- How do I find the test statistic by hand, using a formula (not the calculator)?
- For a two tailed test, how do I find both tails when I only have one test statistic?
- Given the observation, how do I find the test statistic using a formula (not the calculator)?

- In a two tailed test, is the p-value the total shaded area in both tails or just the area in just one tail?
- Given the test statistic, how do I find the observation using algebra (not technology)?

For answers to these *A+ Student Questions* and other hypothesis testing questions, go to allisonlovesmath.com/hypothesis-testing.

CONCLUSION

What do you do next?

Practice.

It's true. To do well on hypothesis testing questions, you must practice. Practice the *Five-Step Crush Hypothesis Testing Method* when you do your homework. Practice showing all five steps every time, even if your teacher doesn't require you to show your work. Practice all five steps until it feels repetitive, obvious, automatic, and . . . easy. If you do, you'll show enough work to get full credit, you'll never miss a step, and you'll *crush* hypothesis testing.

We hope that you found this book helpful and that it answered your questions, clarified a few of the confusing aspects of hypothesis testing, and made your life a little less stressful.

If you found *Crush Hypothesis Testing* helpful, we would appreciate it so much if you would ***leave a review*** for the book in Amazon! It is by far the most impactful way you can support our work and help us to help more students. Thanks a ton!

Warmly,

Allison & Jennifer

ABOUT THE AUTHOR

Allison Dillard is a math professor, host of the Allison Loves Math Podcast, and author of multiple Amazon bestselling books, including Crush Math Now and the Love Math Journal. She has an M.S. in Mathematics from Claremont Graduate University, as well as a B.A. in English and a B.A. in Mathematics from Scripps College. When she is not teaching, writing, or podcasting, you'll find Allison playing board games with her husband, coaching her kids' soccer teams and dreaming up new ways to convince the world that math is awesome. Her love for math is legendary.

facebook.com/AllisonLovesMath

instagram.com/allisonlovesmath

twitter.com/AllisonLuvsMath

ABOUT THE AUTHOR

Jennifer Flenner is a math professor with a PhD from Claremont Graduate University and an extensive career in machine learning and applied statistics. Starting as a civilian scientist working for the Navy, Jennifer applied statistical methods to field testing and earned a distinguished award for most outstanding contribution in the field of Science and Engineering. She has a PhD from Claremont Graduate University (CGU), held a tenure track position at Bakersfield Community College and now, after relocating to Michigan, a tenure track position at Lansing Community College. When Jennifer is not teaching or writing, you can find her biking, cross country skiing, or gardening.

FREE BONUS BOOK

Raise Your Math Grade will teach you the mindset, study habits, and test taking skills necessary to succeed in difficult math classes, like Statistics. Download it for free at:

allisonlovesmath.com/free-book

Made in the USA
Middletown, DE
10 January 2022

58342239R00139